For Attribution—
Developing Data Attribution and Citation Practices and Standards

Summary of an International Workshop

NATIONAL RESEARCH COUNCIL
OF THE NATIONAL ACADEMIES

For Attribution—
Developing Data Attribution and Citation Practices and Standards

Summary of an International Workshop

Paul F. Uhlir, Rapporteur

Board on Research Data and Information

Policy and Global Affairs

NATIONAL RESEARCH COUNCIL
OF THE NATIONAL ACADEMIES

THE NATIONAL ACADEMIES PRESS
Washington, D.C.
www.nap.edu

THE NATIONAL ACADEMIES PRESS   500 Fifth Street, NW   Washington, DC   20001

NOTICE: The project that is the subject of this report was approved by the Governing Board of the National Research Council, whose members are drawn from the councils of the National Academy of Sciences, the National Academy of Engineering, and the Institute of Medicine. The members of the committee responsible for the report were chosen for their special competences and with regard for appropriate balance.

This project was supported by the Alfred P. Sloan Foundation under Grant No. 2011-3-19, and by the Institute of Museum and Library Services under Grant No. 1042078. This report was prepared as an account of work sponsored by an agency of the United States government. Neither the United States government nor any agency thereof, nor any of their employees, makes any warranty, express or implied, or assumes any legal liability or responsibility for the accuracy, completeness, or usefulness of any information, apparatus, product, or process disclosed, or represents that its use would not infringe privately owned rights. Reference herein to any specific commercial product, process, or service by trade name, trademark, manufacturer, or otherwise does not necessarily constitute or imply its endorsement, recommendation, or favoring by the United States government or any agency thereof. Any opinions, findings, conclusions, or recommendations expressed in this publication are those of the authors and do not necessarily reflect the views of the National Academies or the organizations or agencies that provided support for the project.

International Standard Book Number-13: 978-0-309-26728-1
International Standard Book Number-10: 0-309-26728-5

Additional copies of this report are available for sale from the National Academies Press, 500 Fifth Street, NW, Keck 360, Washington, DC 20001; (800) 624-6242 or (202) 334-3313; Internet, http://www.nap.edu/.

Copyright 2012 by the National Academy of Sciences. All rights reserved.

Printed in the United States of America

THE NATIONAL ACADEMIES
Advisers to the Nation on Science, Engineering, and Medicine

The **National Academy of Sciences** is a private, nonprofit, self-perpetuating society of distinguished scholars engaged in scientific and engineering research, dedicated to the furtherance of science and technology and to their use for the general welfare. Upon the authority of the charter granted to it by the Congress in 1863, the Academy has a mandate that requires it to advise the federal government on scientific and technical matters. Dr. Ralph J. Cicerone is president of the National Academy of Sciences.

The **National Academy of Engineering** was established in 1964, under the charter of the National Academy of Sciences, as a parallel organization of outstanding engineers. It is autonomous in its administration and in the selection of its members, sharing with the National Academy of Sciences the responsibility for advising the federal government. The National Academy of Engineering also sponsors engineering programs aimed at meeting national needs, encourages education and research, and recognizes the superior achievements of engineers. Dr. Charles M. Vest is president of the National Academy of Engineering.

The **Institute of Medicine** was established in 1970 by the National Academy of Sciences to secure the services of eminent members of appropriate professions in the examination of policy matters pertaining to the health of the public. The Institute acts under the responsibility given to the National Academy of Sciences by its congressional charter to be an adviser to the federal government and, upon its own initiative, to identify issues of medical care, research, and education. Dr. Harvey V. Fineberg is president of the Institute of Medicine.

The **National Research Council** was organized by the National Academy of Sciences in 1916 to associate the broad community of science and technology with the Academy's purposes of furthering knowledge and advising the federal government. Functioning in accordance with general policies determined by the Academy, the Council has become the principal operating agency of both the National Academy of Sciences and the National Academy of Engineering in providing services to the government, the public, and the scientific and engineering communities. The Council is administered jointly by both Academies and the Institute of Medicine. Dr. Ralph J. Cicerone and Dr. Charles M. Vest are chair and vice chair, respectively, of the National Research Council.

www.national-academies.org

Steering Committee, Developing Data Attribution and Citation Practices and Standards: An International Workshop

Christine Borgman (Chair)
Professor and Presidential Chair
Graduate School of Education and Information Studies
University of California, Los Angeles

Steven Jackson
Assistant Professor, School of Information, and
Director, Technology Policy Culture Research Lab
University of Michigan

Gary King
Albert J. Weatherhead, III. Professor, Department of Government, and
Director, Institute for Quantitative Social Science
Harvard University

David Kochalko
Vice President, Business Strategy and Development, IP & Science
Thomson Reuters

Allen Renear
Associate Dean for Research
University of Illinois at Urbana-Champaign
Graduate School of Library and Information Science

Herbert van de Sompel
Research Scientist
Los Alamos National Lab

John Wilbanks
Vice President, Creative Commons,
Director, Science Commons
Creative Commons

Project Staff at the National Academies

Paul F. Uhlir, Director, Board on Research Data and Information

Daniel Cohen
Program Officer
(on detail from Library of Congress)

Cheryl Williams Levey
Senior Program Associate

BOARD ON RESEARCH DATA AND INFORMATION
MEMBERSHIP (as of the date of this workshop)

Michael Lesk, Chair, Rutgers University

Roberta Balstad, Vice Chair, Columbia University

Maureen Baginski, Serco

Francine Berman, Rensselaer Polytechnic Institute

R. Steven Berry, University of Chicago

Christine Borgman, University of California, Los Angeles

Norman Bradburn, University of Chicago

Bonnie Carroll, Information International Associates

Michael Carroll, American University, Washington College of Law

Paul A. David, Stanford Institute for Economic Policy Department of Economics

Barbara Entwisle, University of North Carolina

Michael Goodchild, University of California, Santa Barbara

Alyssa Goodman, Harvard University

Margaret Hedstrom, University of Michigan

Michael Keller, Stanford University

Michael R. Nelson, Georgetown University

Daniel Reed, Microsoft Research

Cathy H. Wu, University of Delaware and Georgetown University Medical Center

BOARD ON RESEARCH DATA AND INFORMATION
MEMBERSHIP (as of the date of this report)

Francine Berman, Cochair, Rensselaer Polytechnic Institute

Clifford Lynch, Cochair, Coalition for Networked Information

Laura Bartolo, Kent State University

Philip Bourne, University of California, San Diego

Henry Brady, University of California, Berkeley

Mark Brender, GeoEye Foundation

Bonnie Carroll, Information International Associates

Michael Carroll, Washington College of Law, American University

Sayeed Choudhury, Johns Hopkins University

Keith Clarke, University of California, Santa Barbara

Paul David, Stanford Institute for Economic Policy Research

Kelvin Droegemeier, University of Oklahoma

Clifford Duke, Ecological Society of America

Barbara Entwisle, University of North Carolina

Stephen Friend, Sage Bionetworks

Margaret Hedstrom, University of Michigan

Alexa McCray, Harvard Medical School

Alan Title, Lockheed Martin Advanced Technology Center

Ann Wolpert, Massachusetts Institute of Technology

EX OFFICIO

Robert Chen, Columbia University

Michael Clegg, University of California, Irvine

Sara Graves, University of Alabama in Huntsville

John Faundeen, Earth Resources Observation and Science Center

Eric Kihn, National Geophysical Data Center

Chris Lenhardt, Oak Ridge National Laboratory

Kathleen Robinette, Air Force Research Laboratory

Alex de Sherbinin, Columbia University

Board on Research Data and Information Staff

Paul F. Uhlir, Board Director

Subhash Kuvelker, Senior Program Officer

Daniel Cohen, Program Officer (on detail from Library of Congress)

Cheryl Williams Levey, Senior Program Associate

Preface and Acknowledgments

The growth of electronic publishing of literature has created new challenges, such as the need for mechanisms for citing online references in ways that can assure discoverability and retrieval for many years into the future. The growth in online datasets presents related, yet more complex challenges. It depends upon the ability to reliably identify, locate, access, interpret and verify the version, integrity, and provenance of digital datasets.

Data citation standards and good practices can form the basis for increased incentives, recognition, and rewards for scientific data activities that in many cases are currently lacking in many fields of research. The rapidly-expanding universe of online digital data holds the promise of allowing peer-examination and review of conclusions or analysis based on experimental or observational data, the integration of data into new forms of scholarly publishing, and the ability for subsequent users to make new and unforeseen uses and analyses of the same data – either in isolation, or in combination with other datasets.

The problem of citing online data is complicated by the lack of established practices for referring to portions or subsets of data. As funding sources for scientific research have begun to require data management plans as part of their selection and approval processes, it is important that the necessary standards, incentives, and conventions to support data citation, preservation, and accessibility be put into place.

There are, in fact, a number of initiatives in different organizations, countries, and disciplines already underway. An important set of technical and policy approaches have already been launched by the U.S. National Information Standards Organization (NISO) and other standards bodies regarding persistent identifiers and online linking. Another important group is DataCite. The World Data System is also focusing on these issues, but other initiatives remain ad hoc and uncoordinated.

The workshop summarized here was organized by a steering committee under the National Research Council's (NRC's) Board on Research Data and Information, in collaboration with an international CODATA-ICSTI Task Group on Data Citation Standards and Practices. The purpose of the symposium was to examine a number of key issues related to data identification, attribution, citation and linking, to help coordinate activities in this area internationally, and to promote common practices and standards in the scientific community. More specifically, the statement of task for this project asked the following questions:

1. What is the status of data attribution and citation practices in the natural and social (economic and political) sciences in United States and internationally?

2. Why is the attribution and citation of scientific data important and for what types of data? Is there substantial variation among disciplines?

3. What are the major scientific, technical, institutional, economic, legal, and socio-cultural issues that need to be considered in developing and implementing scientific data citation

standards and practices? Which ones are universal for all types of research and which ones are field or context specific?

4. What are some of the options for the successful development and implementation of scientific data citation practices and standards, both across the natural and social sciences and in major contexts of research?

The workshop that was organized pursuant to these questions was held in Berkeley, CA on August 22-23, 2011. The presentations and discussions that are summarized from this meeting in the volume that follows are part of this effort.

This report has been prepared by the workshop rapporteur as a factual summary of what occurred at the workshop. The committee's role was limited to planning and convening the workshop. The views contained in the report are those of the individual workshop participants and do not necessarily represent the views of all workshop participants, the planning committee, or the National Academies.

Acknowledgments

We are grateful to the following for support of this project: Institute of Museum and Library Services, grant number IMLS LG-00-11-0123-11; Sloan Foundation, grant number 2011-3-19; the Committee on Data for Science and Technology (CODATA); and Microsoft Research. Any views, findings, conclusions or recommendations expressed in this publication do not necessarily represent those of the Institute of Museum and Library Services, or the other sponsors.

This report has been reviewed in draft form by individuals chosen for their diverse perspectives and technical expertise, in accordance with procedures approved by the National Academies' Report Review Committee. The purpose of this independent review is to provide candid and critical comments that will assist the institution in making its published report as sound as possible and to ensure that the report meets institutional standards for quality and objectivity. The review comments and draft manuscript remain confidential to protect the integrity of the process.

We wish to thank the following individuals for their review of this report: Suzanne Allard, University of Tennessee; Anne Fitzgerald, Queensland University, Australia; Charles Humphrey, University of Alberta; Brian McMahon, International Union of Crystallography, United Kingdom; and John Rumble, Information International Associates (retired).

Although the reviewers listed above have provided many constructive comments and suggestions, they were not asked to endorse the content of the report, nor did they see the final draft before its release. Responsibility for the final content of this report rests entirely with the rapporteur and the institution.

Many people devoted many months of effort to organizing this event. Dan Cohen and Cheryl Levey of the staff of the Board on Research Data and Infrastructure spent much of their 2011 summer working on the Workshop project. Christine Borgman, Paul Uhlir, and Dan Cohen had conference calls with each session panel to ensure synthesis and continuity. The Workshop was coordinated with the activities of the CODATA-ICSTI Task Group on Data Citation Standards and Practices, whose co-chairs are Bonnie Carroll, Jan Brase, and Sarah Callaghan. Members of that Task Group are (in alphabetical order) Micah Altman, Elisabeth Arnaud, Christine Borgman, Dora Ann Lange Canhos, Todd Carpenter, Vishwas Chavan, Michael Diepenbroek, John Helly, Jianhui Li, Brian McMahon, Karen Morgenroth, Yasuhiro Murayama, Helge Sagen, Eefke Smit, Martie van Deventer, John Wilbanks, and Koji Zettsu. Paul Uhlir, Dan Cohen, and Franciel Linares are staff consultants to the Task Group. Special thanks also are due to the Workshop Steering Committee, consisting of Christine Borgman (Chair), Allen Renear, Herbert van de Sompel, Gary King, Steven Jackson, David Kochalko, and John Wilbanks, as well as to the young scientists who served as rapporteurs in the final afternoon sessions: Franciel Linares, Matthew Mayernick, Jillian Wallis, and Laura Wynholds.

Christine Borgman
Steering Committee Chair

Paul F. Uhlir
Project Director

Contents

1- Why Are the Attribution and Citation of Scientific Data Important? 1
    Christine Borgman

PART ONE - TECHNICAL CONSIDERATIONS 9

2- Formal Publication of Data: An Idea Whose Time Has Come? 11
    Jean-Bernard Minster

3- Attribution and Credit: Beyond Print and Citations 15
    Johan Bollen

4- Data Citation—Technical Issues —Identification 23
    Herbert Van de Sompel

5- Maintaining the Scholarly Value Chain: Authenticity, Provenance, and Trust 31
    Paul Groth

DISCUSSION BY WORKSHOP PARTICIPANTS 35
    Moderated by John Wilbanks

PART TWO - DISCIPLINE-SPECIFIC ISSUES 41

6- Towards Data Attribution and Citation in the Life Sciences 43
    Philip Bourne

7- Data Citation in the Earth and Physical Sciences 49
    Sarah Callaghan

8- Data Citation for the Social Sciences 55
    Mary Vardigan

9- Data Citation in the Humanities: What's the Problem? 59
    Michael Sperberg-McQueen

DISCUSSION BY WORKSHOP PARTICIPANTS 65
    Moderated by Herbert van de Sompel

PART THREE - LEGAL, INSTITUTIONAL, AND SOCIO-CULTURAL ASPECTS 69

10- Three Legal Mechanisms for Sharing Data 71
    Sarah Hinchliff Pearson

11- Institutional Perspective on Credit Systems for Research Data 77
    MacKenzie Smith

12- Issues of Time, Credit, and Peer Review 81
    Diane Harley

DISCUSSION BY WORKSHOP PARTICIPANTS 89
  Moderated by Paul F. Uhlir

PART FOUR - EXAMPLES OF DATA CITATION INTITIATIVES 93

13- The DataCite Consortium 95
  Jan Brase

14- Data Citation in the Dataverse Network ® 99
  Micah Altman

15- Microsoft Academic Search: An Overview and Future Directions 107
  Lee Dirks

16- Data Center-Library Cooperation in Data Publication in Ocean Science 109
  Roy Lowry

17- Data Citation Mechanism and Service for Scientific Data: Defining a Framework for Biodiversity Data Publishers 113
  Vishwas Chavan

18- How to Cite an Earth Science Dataset? 117
  Mark Parsons

19- Citable Publications of Scientific Data 125
  John Helly

20- The SageCite Project 131
  Monica Duke

DISCUSSION BY WORKSHOP PARTICIPANTS 137
  Moderated by David Kochalko

PART FIVE - INSTITUTIONAL PERSPECTIVES 141

21- Developing Data Attribution and Citation Practices and Standards: An Academic Institution Perspective 143
  Deborah L. Crawford

22- Data Citation and Data Attribution: A View from the Data Center Perspective 147
  Bruce E. Wilson

23- Roles for Libraries in Data Citation 151
  Michael Witt

24- Linking Data to Publications: Towards the Execution of Papers 157
  Anita De Waard

25- Linking, Finding, and Citing Data in Astronomy 161
  Michael J. Kurtz

DISCUSSION BY WORKSHOP PARTICIPANTS 167
    Moderated by Bonnie Carroll

26- Standards and Data Citations 173
    Todd Carpenter

27- Data Citation and Attribution: A Funder's Perspective 177
    Sylvia Spengler

DISCUSSSION BY WORKSHOP PARTICIPANTS 179
    Moderated by Christine Borgman

PART SIX SUMMARY OF BREAKOUT SESSIONS 187

Breakout Session on Technical Issues 189
    Moderator: Martie van Deventer
    Rapporteur: Franciel Linares

Breakout Session on Scientific Issues 193
    Moderator: Sarah Callaghan
    Rapporteur: Matthew Mayernik

Breakout Session on Institutional, Financial, Legal, and Socio-cultural Issues 199
    Moderator: Vishwas Chavan
    Rapporteur: Laura Wynholds

Breakout Session on Institutional Roles and Perspectives 209
    Moderator: Bonnie Carroll
    Rapporteur: Jillian Wallis

Appendix A: Agenda 211

Appendix B: Speaker and Moderator Biographical Information 217

Why Are the Attribution and Citation of Scientific Data Important?

Christine Borgman[1]
University of California at Los Angeles

Introduction

My roles as the project Chair and as keynoter are to frame the problems to be addressed in two days of discussion. This is a very sophisticated set of speakers and participants. Each of you has been concerned with research data, in some way, for some years. By now, all of us are familiar with the data deluge metaphor. We are being drowned in data, much of which is runoff. Valuable research data often are not captured, cited, or reused. Our challenge is to identify what part of these resources should be kept, the right way to keep them, and the right tools and services to make them useful.

Data have become a critical focus for scholarly communication, information management, and research policy. We cannot address the full array of these issues, fascinating though they may be. Our two days will focus closely on questions of attribution and citation of scientific research data, although we frame *scientific* broadly enough to include most areas of scholarship.

We will devote little time to definitional issues (e.g., what are data?), except to acknowledge that *data* often exist in the eyes of the beholder. Our principal concerns are how to assign credit for data (attribution) and how to reference data (citation) in ways that others can identify, discover, and retrieve them. Among the questions to be explored are what a community considers to be data, what data might be shared, what data should be shared, when data can be shared, and in what forms can data be shared? We will consider what approaches may be generic across disciplines and what practices may be field-specific.

Data citation and attribution are not new topics.[2] We have had standards for cataloging data files since the 1970s. Objects that can be cataloged also can be cited. Similarly, data archives have been promoting data citation practices for several decades. However, over this same period, very few journal editors required data citations, disciplines did not instill data citation as a fundamental practice of good research, granting agencies did not reward the data citations of applicants, tenure and reward committees did not recognize data citations in annual performance reviews, and researchers did not take responsibility for citing data sources. What have we learned from the past? What seems to be new today?

Several developments contribute to the renewed interest in data citation and attribution, all of which are topics of this Workshop. One is the growth in data volume relative to storage and analytic capacities. Fields such as astronomy, physics, and genomics are producing more data than investigators can investigate themselves. By sharing and combining data from multiple sources, other researchers can ask new questions. Another factor is advances in the technical

---

[1] Presentation slides are available at: http://sites.nationalacademies.org/PGA/brdi/PGA_064019.
[2] Thanks to an anonymous reviewer for suggesting a fuller discussion of drivers for data citation and attribution than was included in the oral presentation at the meeting. Some of the reviewer's comments are included in this text.

infrastructure for generating, managing, analyzing, and distributing data. Tools are more sophisticated, bandwidth capacity is greater, and transfer speeds continue to improve. Third, and by no means least, are associated shifts in research policy. Data are now viewed as significant research products in themselves, more than just adjuncts to publications.[3] Funding agencies now expect investigators to capture, manage, and share their data. When viewed as research products, data deserve attribution similar to that of publications. Attribution, in turn, requires mechanisms for references to be made and citations to be received. Yet data are very different entities than publications. They take many more forms, both physical and digital, are far more malleable than publications, and practices vary immensely by individual, by research team, and by research area. Institutional practices to assure stewardship of data are far less mature than are practices to sustain access to publications. All of these factors contribute to the complexity of data citation and attribution. It is the many interacting dimensions of data attribution and citation that make it a problem worthy of this Workshop and of the multi-year effort with which the meeting is associated.

Scholarly Infrastructure

Questions of data citation and attribution are best framed in terms of the infrastructure for digital objects. For our purposes, scholarly infrastructure is captured by the eight dimensions of infrastructure identified by Susan Leigh Star and Karen Ruhleder (1996)[4], as mapped in Figure 1-1, taken from Bowker, et al (2010)[5]:

---

[3] Borgman, C. L. (2012). The conundrum of sharing research data. Journal of the American Society for Information Science and Technology, 63(6): 1059-1078. http://dx.doi.org/10.1002/asi.22634.

[4] Star, S. L. & Ruhleder, K. (1996). Steps toward an ecology of infrastructure: Design and access for large information spaces. Information Systems Research, 7(1): 111-134.

[5] Bowker, G. C., Baker, K., Millerand, F., Ribes, D., Hunsinger, J., Klastrup, L. & Allen, M. (2010). Toward Information Infrastructure Studies: Ways of Knowing in a Networked Environment. In Hunsinger, J., Klastrup, L. & Allen, M. (Eds.). International Handbook of Internet Research. Dordrecht, Springer Netherlands: 97-117.

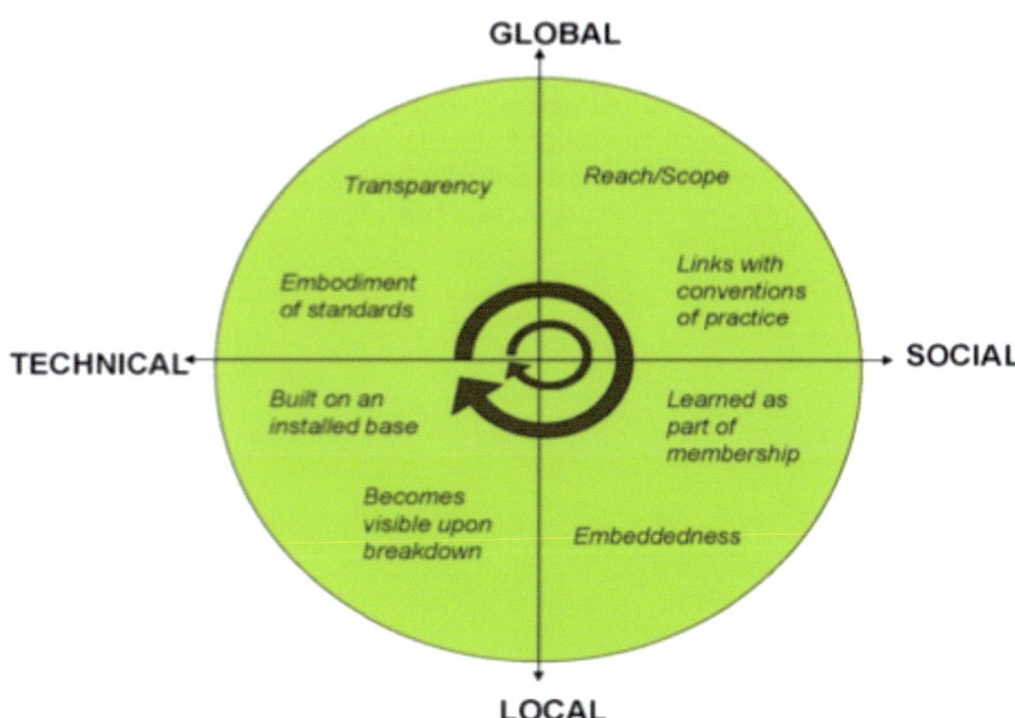

FIGURE 1-1 Dimensions of Infrastructure.
SOURCE: Bowker, G. C., Baker, K., Millerand, F., Ribes, D., Hunsinger, J., Klastrup, L. & Allen, M. (2010). Toward Information Infrastructure Studies: Ways of Knowing in a Networked Environment. In Hunsinger, J., Klastrup, L. & Allen, M. (Eds.). International Handbook of Internet Research. Dordrecht, Springer Netherlands: 97-117.

Our presentations will touch upon each part of this model. At the technical edge of the model, infrastructure is the embodiment of standards, which in turn are built on an installed base. Among the installed bases that influence data citation are Internet protocols, publishing practices, and library cataloging methods. At the social end, infrastructure is linked to conventions of practice–whether cataloging or data management–and learned as part of membership in a community (e.g., librarians or astronomers). A social topic of particular interest is the relationship of reward systems to data citation. At the local edge of the model are individual practices for managing data and library practices for data stewardship. The global edge represents the inherently international character of scientific scholarship. Data practices, data exchange, and citation and attribution all must work effectively across political, institutional, and disciplinary boundaries.

Data in the global-technical quadrant of Figure 1-1 are most amenable to automated capture, management, and discovery. These are data, for example, from shared instruments such as space-based telescopes, and are associated with established data structures, analytical tools, and repositories. It is these types of data that are most readily cited. Conversely, data in the local-social quadrant tend to be more heterogeneous in form and content, more artisanal in data collection methods, and more varied in practices for management, use, and reuse. These data are much less amenable to established methods of data ingest, stewardship, citation, and attribution.

The infrastructure for digital objects has many features. We are concerned at this Workshop with how they apply to data attribution and citation, but we must remember that they are part of a larger Internet architecture of digital objects. The list of features below, around which the rest of my presentation is organized, is neither exhaustive nor mutually exclusive. Rather, it is a useful starting point to assess how these infrastructure features are applied to data citation and attribution:

- Social practice
- Usability
- Identity
- Persistence
- Discoverability
- Provenance
- Relationships
- Intellectual property
- Policy

Social practice

Among the drivers for this Workshop are: renewed interest in data citation due to increases in data volume, to advances in technical infrastructure, and to shifts in research policy associated with data. These developments still beg the questions of why data should be attributed and cited. Those questions have at least as many answers as there are persons attending this event. At the highest level, most of these answers can be grouped into categories of reproducing research, replicating findings, or more generally, to reuse data. To reuse data, it is necessary to determine what useful data exist, where, in what form, and how to get them. In turn, data must be described in some way if they are to be discoverable. For people to invest effort in making data discoverable, they should receive credit for creating, cleaning, analyzing, sharing, and otherwise making data available and useful. To get credit, some means must exist to associate names of individuals and organizations with specific units of data.

This project is titled with the awkward phrase "Developing Data Attribution and Citation Practices and Standards" to make the point that these are not equivalent concepts. The distinction is both subtle and important. Attribution is made to the responsible party. Attribution might thus be given to an individual investigator, to a research team, to a university, to a funding agency, to a data repository, to a library, or to another party responsible for gathering, assembling, curating, or otherwise contributing to the availability of data for others to use. Attribution is more closely associated with the notion of contribution, or contributor, than with author, which is among the differences between handling data and handling publications. Like publications, however, attribution implies social responsibility to give credit where credit is due. When we write journal articles and books, we reference other publications and the evidence on which they are based to attribute our sources.

Citation, in contrast, is the mechanism by which one makes references to other entities. In bibliometric parlance, references are made and citations are received. Even in the bibliographic world, reference/citation formats are many and varied: the *American Psychological Association* standard is popular in the social sciences, the *Modern Language Association* standard in the humanities, the *Association for Computing Machinery* in computer science and engineering, and the *Blue Book* in law, for example. These standards vary by the units they reference (e.g., full publications or individual pages), presentation (e.g., numerical references to the bibliography or author names and dates in text), choice of data elements (e.g., author, title, date, volume, issue, page numbers, DOI, URL, legal jurisdiction), and other factors. The multiplicity of bibliographic standards reflects the diversity of practices within and between research areas. None of them map easily to data or datasets, which have yet more diversity in form and practice.

Usability

Data citation and attribution must be considered in the context of the usability of data as digital objects. While data in the form of physical objects (e.g., samples, artifacts, lab notebooks on paper) also must be referenced, digital descriptions of those objects typically serve as surrogates. Among the actions people – or machines – may wish to perform on digital objects are to interpret, evaluate, open, read, compute upon, reuse, combine, describe, and annotate. This incomplete list suggests the range of capabilities that must be accommodated by a successful system for citing and attributing data.

Identity

To be citable and attributable, data must be identified uniquely. Identity and identification are well known problems in computer science and in epistemology. Our speakers on these topics bring those fields to bear on the question of identity for units of data. Identity is complex when we think in terms of people reading books and reading data. Humans can disambiguate similar objects, such as different editions of a book. Identity questions are even more complex when computers are discovering, reading, and interpreting data. Identity also is closely intertwined with usability and with trust. Among the questions to ask are: What are the dimensions of data identity? What identity levels are necessary to open, to interpret, to read, to compute upon, to combine, and to trust data as digital objects? An effective set of identity mechanisms for data citation and attribution must incorporate a trust fabric.

Persistence

The next session in this Workshop is on identity and persistence of digital objects. Identity and persistence tend to be more concerned with containers of the data than about the data per se – how we package, name, and label data will influence the ability to identify them, to ensure they persist, or to dispose of them accurately. The data may exist, but unless we have labeled them and stored them in a place to which others can return, their usability will be negatively affected. A variety of persistent identifier systems already exist, including Uniform Resource Identifiers (URI), Digital Objects Identifiers (DOIs) and other types of Handles, and other namespaces. While all are useful, none addresses all of the needs for data identity and persistence. Much remains to be learned about which systems are best, for which types of data, and for what purposes.

Discoverability

Discoverability is a broad topic, most researched in information retrieval. For the purposes of data citation and attribution, discoverability is the ability to determine the existence of a set of data objects with specified attributes or characteristics. The attributes of interest include the producer of the data, the date of production, the method of production, a description of an object's contents, and its representation. Discoverability may also include aspects such as levels of quality, certification, and validation by third parties. Discoverability depends both on the description and representation of data and on tools and services to search for data objects. Description and representation usually take the form of metadata, some of which may be automated if data are generated by instruments such as sensor networks or telescopes. Even for these types of data, metadata creation may require considerable human effort, making it an expensive process that is often avoided by researchers.

Human intervention is necessary to add metadata and description to most other kinds of data. As data move from one place to the next, those metadata may be augmented incrementally. Unlabeled bits are equivalent to books shorn of their covers and title pages. Data generally are discoverable via the metadata that describe them.

A variety of approaches to discovery are possible. Web search engines are one possibility, assuming that data descriptions are reachable via standard web protocols. With the introduction of semantic web technologies and associated search engines, location of datasets of interest based on semantic content becomes possible. Alternatively, more discipline-specific and structured catalogs can be created.

Data are discoverable only as long as someone keeps them, somewhere. Library and archival practice tends toward saving forever anything that is worth saving, although both professions also have long histories of weeding collections and of scheduling record disposal. Individual investigators are less likely, and less able, to maintain data permanently for discovery at some unknown later date.

Discoverability is thus associated with economics, a topic largely beyond the scope of this meeting. Many research libraries and archives view data as important special collections, but also are concerned that data stewardship is an unfunded mandate. Data retention schedules will influence data discovery. Some data may be discoverable only in the short term, such as a scratch space for other people to use. Other data will be kept at least until the associated reports are published, and for some time thereafter. Yet other data will be "long-lived," usually defined long enough to be concerned about migration from one format to the next[6,7]. Data citation and attribution practices and standards may vary considerably depending on the period of time data are expected to remain available. Considerations also will vary between raw data, observations, models, physical samples, the predicted life span of utility, and many other factors.

---

[6] Reference Model for an Open Archival Information System (2002). Recommendation for Space Data System Standards: Consultative Committee for Space Data Systems Secretariat, Program Integration Division (Code M-3), National Aeronautics and Space Administration. http://public.ccsds.org/publications/archive/650x0b1.pdf.

[7] Long-Lived Digital Data Collections. (2005). National Science Board. http://www.nsf.gov/pubs/2005/nsb0540/.

Provenance

Provenance is particularly important for data citation. In citing data, it is important to reference the correct version, and where possible, to cite prior states of data and the transformations made between states. Provenance was once the exclusive concern of museum curators, archivists, and forensic specialists, all of whom view provenance as the chain of custody of an object. For example, The Getty Museum trusts the authenticity of an artwork only if the custody of that object can be documented at all steps since its origin. This linear model of provenance is less applicable to digital objects. In computing, provenance is the ability to track all transformations from the original state. Data provenance is becoming an active research area, and one to which we devote a substantial time at this Workshop.

Relationships

While data can be discrete digital objects, they usually are related to other objects such as publications. Often multiple types of data have relationships to each other, providing context, calibration, and comparisons. Data citation and attribution mechanisms thus must facilitate linking of related objects and be able to refer to groups of objects, as well as to individual items. The choice of units for reference is a particularly contentious topic in data citation and attribution. When does citing a dataset associated with a journal article provide sufficient granularity? When is it necessary to cite each observation, each cell in a table, or each point on a graph? Identifying units, relationships among units, and types of relationships are all aspects of data citation and attribution.

Intellectual property

Intellectual property is a broad topic even if confined to scientific data. The discussion at this meeting on intellectual property will focus on rights associated with data, such as the rights to use, reuse, combine, publish, and republish. Discovering data is but a first step. Once discovered and retrieved, users need to be able to identify what rights are associated with those data. For example, may we use the data for commercial purposes? May we share them with others? May we use them for teaching? Research teams, especially small teams, may not have documented ownership or rights associated with their data. Until data came to be viewed as valuable research products, ownership was an issue rarely discussed. Data often are not shared for the simple reason that it is not possible to determine who in a collaborative project has the rights to release them.

Open access, albeit an overused term with many meanings, has sensitized researchers to the value of making their research products available. From an intellectual property perspective, making a reference to data should be no different than a reference to a book or published paper. Including bibliographies in published works does not violate the copyright of the works cited. Rather, the bibliography is the form of attribution most central to scholarly practice.

Policy

Both data citation and attribution have policy components. Many stakeholders are concerned with scholarly information policy, including funding agencies, publishers, data repositories, universities, investigators, and students. Each has policy concerns, thus we must ask what policy, what kinds of policy, and whose policy? Data management plan requirements and data sharing policies are case examples. Many funding agencies have established such policies, the specifics of which vary widely between the National Science Foundation and the National Institutes of Health in the United States, the Wellcome Trust and the Economic and Social Research Council in the United Kingdom, and others in the U. K., the European Union, and Asia. These requirements may evolve to become more explicit about who is to receive what kinds of attribution for what kinds of data contributions, and how such contributions are to be cited.

Workshop themes

All of these infrastructure issues, and more, will be explored in our program. The two days of the Workshop are organized around these driving questions from the project's task statement (emphasis added):

1. What are the **major technical issues** that need to be considered in developing and implementing scientific data citation standards and practices?

2. What are the **major scientific issues** that need to be considered in developing and implementing scientific data citation standards and practices? Which ones are **universal** for all types of research and which ones are **field- or context- specific**?

3. What are the **major institutional, financial, legal, and socio-cultural** issues that need to be considered in developing and implementing scientific data citation standards and practices? Which ones are universal for all types of research and which ones are field- or context-specific?

4. What is the **status of data attribution and citation practices** in individual fields in the natural and social (economic and political) sciences in the United States and internationally? Provide case studies.

5. **Institutional Roles and Perspectives:** What are the respective roles and approaches of the main actors in the research enterprise and what are the similarities and differences in disciplines and countries? The roles of research funders, universities, data centers, libraries, scientific societies, and publishers will be explored.

Next steps

This summary report of the Workshop, published by the National Academy of Sciences' Board on Research Data and Information, is the first formal product of the overall initiative on data attribution and citation. The CODATA-ICSTI Task Group on Data Citation Standards and Practices is conducting a survey and literature review, and gathering other materials for an international white paper. Task Group members are giving presentations at international meetings over the next several years. Future efforts of this Task Group are expected to lead to standardization work. These efforts also will continue via the participants' dissemination of the ideas generated here.

PART ONE

TECHNICAL CONSIDERATIONS

2- Formal Publication of Data: An Idea Whose Time Has Come?

Jean-Bernard Minster[1]
University of California at San Diego

Every time I participate in a discussion on data citation and attribution or talk to colleagues who deal with a lot of data, the issue of data publication comes up. The point is that the whole idea of citation is difficult to discuss in the absence of the concept of publication. The idea of long-term data preservation, citation, and publication is a concept that is growing in the community. In my scientific union, the American Geophysical Union (AGU), there is a statement on data publication that reads:

> The cost of collecting, processing, validating, and submitting data to a recognized archive should be an integral part of research and operational programs. Such archives should be adequately supported with long-term funding. Organizations and individuals charged with coping with the explosive growth of Earth and space digital data sets should develop and offer tools to permit fast discovery and efficient extraction of online data, manually and automatically, thereby increasing their user base. The scientific community should recognize the professional value of such activities by endorsing the concept of publication of data, to be credited and cited like the products of any other scientific activity, and encouraging peer-review of such publications.[2]

If you look at the literature, Figure 2-1 from the paper by Hilbert and Lopez shows the growth in total information. What is amazing is that between 1986 and 2007, everything having to do with vinyl (analog) has disappeared and everything that is now digital becomes completely dominated by PCs. Most of the data we have now are on people's PCs. The growth has been quite constant. So, for any conclusion that you draw from a study from 1986 to 2007, you probably have to scale those estimates upward considerably, in order to assess the situation today accurately.

---

[1] Presentation slides are vailable at http://sites.nationalacademies.org/PGA/brdi/PGA_064019.
[2] "The Importance of Long-term Preservation and Accessibility of Geophysical Data" AGU, May 2009.

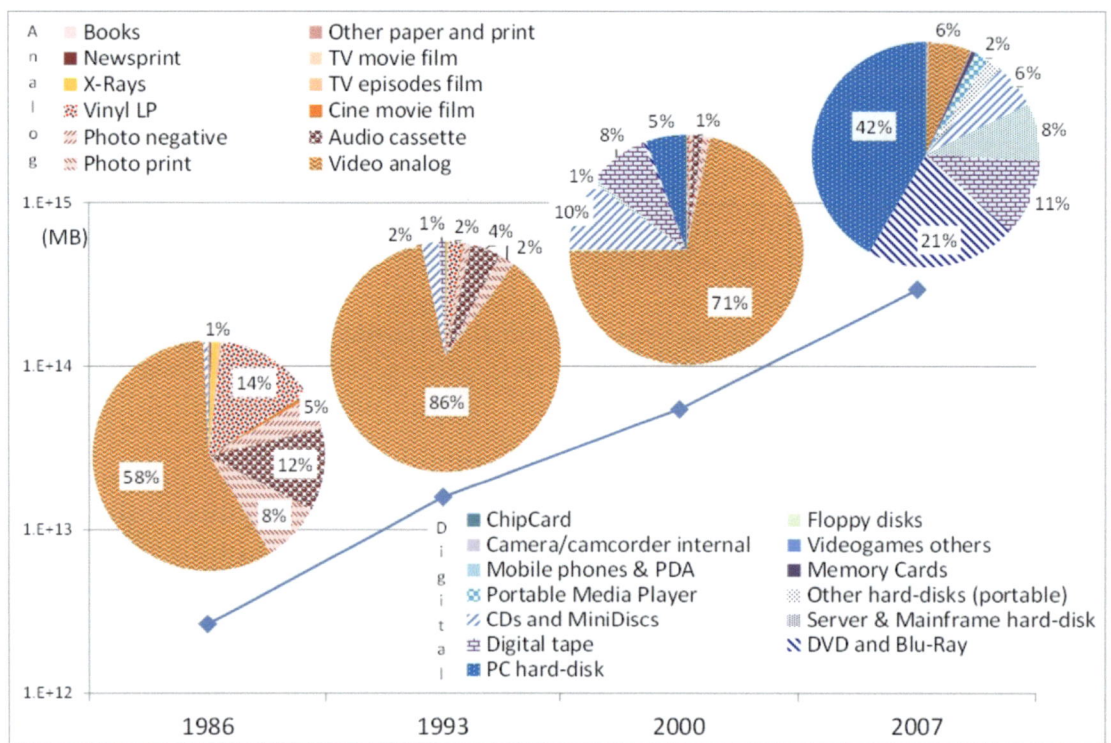

FIGURE 2-1 Growth in total information.
SOURCE: Hilbert, M. & Lopez, P. (2011). The World's Technological Capacity to Store, Communicate, and Compute Information. Science, 332, 60-65 [doi: 10.1126/science.1200970].

The storage capacity globally is shrinking in relation to amount of information. Data compression technologies show some promise in addressing the mismatch between our growing storage needs and the available capacities. Lossless[3] compression strategies have already been deployed in many data centers, but the compression ratios they can provide are fairly modest for many types of data. Existing lossy compression methods (i.e., compression algorithms that achieve greater compression ratios at the cost of some degradation to the quality of the original data), such as those now available for digital images and video, are problematic for some kinds of data because we do not know what information may be important to future researchers.

---

[3] Lossless compression algorithms restore 100 percent of the original data upon decompression. They achieve compression by techniques such as representing strings of repeated instances of the same character with a single instance plus additional characters indicating the number of repetitions in the original. The files compressed using lossless compression may require the use of a decompression algorithm in order to be read by the application that created them, but once decompressed, the resulting file is essentially identical to the original. Lossy compression algorithms achieve greater compression ratios at the cost of the loss of some portion of the information contained in the original. For example, lossy compression techniques may reduce the color depth or resolution of a graphical image, may use a lower sampling rate of audio content, or may preserve only the delta between frames of a video sequence rather than the entirety of all the frames. The compressed result is an approximation of the original that is "good enough" for many purposes, but once so compressed, the content cannot be restored to the quality of the original prior to compression. For some types of content, lossy compression techniques can achieve dramatically higher compression ratios than lossless techniques, but carry the risk that something lost in the compression process may be important for a future use perhaps not contemplated at the time.

Consequently, data center managers are reluctant to use these methods and continue to rely instead upon the continued expansion of physical storage capacity.

We have to find a way of saving the materials that are worth saving and this can be achieved through the process of publication. We all have enormous file cabinets in our offices, but the information that is published is really what gets preserved for a long time. The problem of how to deal with the growing deficit in storage capacity is beyond the scope of this workshop, but it is worth noting that citation to data has little value if the data being cited are not preserved and accessible for however long they may be needed.

This whole idea of data publication, citation, and attribution is a very current concept. However, some best practices and critical research needs are beginning to emerge. It is also getting increasing attention from the scientific community. For example, there was a whole session on these topics at the CODATA conference in October 2010 in Cape Town, South Africa. Also, another session will be devoted to these issues at the World Data Systems science symposium in Kyoto, Japan in September 2011. The International Council for Science (ICSU) envisions a global World Data System (WDS) that will:

- Emphasize the critical importance of data in global science activities,
- Further ICSU strategic scientific outcomes by addressing pressing societal needs (e.g., sustainable development, the digital divide).
- Highlight the very positive impact of universal and equitable access to data and information.
- Support services for long-term stewardship of data and information.
- Promote and support data publication and citation.

The maturity of the development of these practices is not uniform across fields and disciplines, however. In crystallography, for example, you do not get credit for your work unless you publish your data and it has to be published in certain formats. The field has procedures and protocols. This is an example of a discipline that is very well organized. It is not the same in other fields, although the technology is available.

The WDS faces certain challenges however In order to accommodate at the same time giant data facilities, such as the NASA Distributed Active Archive Centers or NOAA National Data Centers, and very small facilities such as the WDC for Earth Tides, the same model will not work equally well. Similarly, the International Global Navigation Satellite System, which involves an enormous projected data flow, will function according to a certain model, but the very small international data services, such as those for the glaciological or the solar physics communities for example, will function in a very different way.

Not all WDS members are capable of providing all the necessary infrastructure components identified here. Consequently, the WDS Scientific Committee realized that one type of membership was inadequate. It created four separate types of memberships, described in some detail on the WDS website. So far, only "regular" members have the mandate to provide a "secure repository" function. However, the definition of WDS member roles is still a work in progress.

So what is the purpose of data citation? It is, as I see it, to give credit and make authors accountable, and to aid in the reproducibility of science. This is a way we could cite data:

> Cline, D., R. Armstrong, R. Davis, K. Elder, and G. Liston. 2002, Updated July 2004. CLPX-Ground: ISA snow pit measurements. Edited by M. Parsons and M. J. Brodzik. Boulder, CO: National Snow and Ice Data Center. Data set accessed 2008-05-14 at http://nsidc.org/data/nsidc-0176.html.

In this example, we have a description of a dataset. It shows the proper citation of certain data out of the total number of entries, who is responsible for the dataset, who edited it, what was the location, and when it was last accessed online. The latter element may be important for some continuously changing datasets (e.g., time-series weather records); it is often much less important than a specific version number or revision date of the dataset. This, of course, assumes that the data "publisher" both maintains a clear history and can provide access to specific revisions of the dataset.

While it is not difficult to specify these elements for a data citation, even this fairly simple citation format received negative feedback from some researchers in my field. Some of my colleagues and students said: "We cannot possibility remember all those things. It is just too hard." This suggests that, at least in some fields and disciplines, the cultural challenges may be greater than the technical ones.

Let me conclude with what I think is needed:

- Data collection coupled with quality control
- Quality assurance (a function of the data)
- Peer review ascertaining the authoritative source, assessed data
- Ease of publication
- Easily understood standards (especially metadata)
- Simple steps to place data in the public domain (e.g., the Polar Information Commons)
- Secure repository and long-term data curation
- Preferred use of this reliable source by data users
- Preservation of long-term data time series
- Repositories that adapt to evolving technology
- Collaboration with libraries and the publishing communities
- Ease of citation
- Credit given to data authors and proper recognition and citation by users
- Professional recognition (besides credit)
- Perhaps a change in academic mind-set.

3- Attribution and Credit: Beyond Print and Citations

Johan Bollen[1]
Indiana University

The main focus of my work is not on citations, let alone data citations, but on computational methods to study scientific communication by analyzing very large-scale usage data. This is quite different from citation data, but how we organize and analyze our data is probably a useful and worthwhile perspective to contribute here.

When researchers talk about data citations, the assumption is that a citation has value. But why is it valuable? It is valuable because it defines a notion of credit and attribution in scientific communication. It is the mechanism by which one author explicitly indicates that he or she has been influenced by the thinking or the work of another author. Citations are very strongly grounded in the tradition of printed scientific paper, but we are thinking about data now, and data is much more difficult to cite in that context. The main problem here is that technology has fundamentally changed scholarly communication, and in fact even how scholars think, but scholarly review and assessment are still stuck in the paper era (e.g., peer review, print, citations, journals) that we have known since the late 19th century. However, if you look at how scholarly communication has been evolving over just the past 10 to 15 years, most of it has moved online. Most of my colleagues are on Twitter and Facebook now. One of the ways that they communicate their science is by posting tweets that make references to their papers and data. In other words, the way they publish has fundamentally changed.

This is also true for my own experience. When I write a paper the first thing I do is to deposit it in my web site or in an archive. The community then finds its way to my paper and if people find errors, they will provide extensive feedback, through, for example, a blog post. So, in addition to publishing my papers online, they are also "peer reviewed" online. The whole notion here is that the entire spectrum of scholarly communication is moving online. Before, it seemed to be occurring mostly within the confines of the traditional publication system.

If you look at scholarly assessment, however, it seems like it has skipped that evolution nearly entirely. Therefore, I think that we need to talk about changing scholarly assessment beyond the traditional way of doing things, to systems that can actually keep up with the changes in the scholarly communication process.

Figure 3-1 shows that publication data and citation data are the end-product of a long chain of scholarly activities. Usage data can be harvested for each of the antecedent activities, such as when authors read the scholarly literature as part of their research, submission, and peer-review process.

---

[1] Presentation slides are available at http://sites.nationalacademies.org/PGA/brdi/PGA_064019.

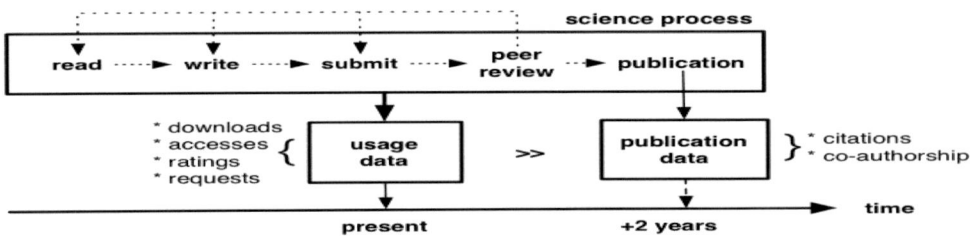

FIGURE 3-1 Data for assessing scholarly communication.

For that reason we have looked at applications of usage data for scholarly assessment. The main promise of usage data is that it can be recorded for all digital scholarly content (e.g., papers, journals, preprints, blog postings, data, chemical structures, software), not just for 10,000 journals and not only for peer-reviewed, scholarly articles. It provides extensive information on types of user behavior, sequences, timing, context, and clickstreams. It also reflects the behavior of all users of scholarly information (e.g., students, practitioners, and scholars in domains with different citation practices). Furthermore, interactions are recorded starting immediately after publication; that is, the data can reflect real-time changes (see Figure 3-1). Finally, usage data offers very large-scale indicators of relevance—billions of interactions recorded for millions of users by tens of thousands of scholarly communication services.

However, there are significant challenges with usage data. These include:

(1) Representativeness: usage data is generally recorded by a particular service for its particular user community. To make usage data representative of the general scholarly community, i.e. beyond the user base of a single service, we must find ways to aggregate usage data across many different services and user communities.

(2) Attribution and credit: a citation is an explicit, intentional expression of influence, i.e., authors are explicitly acknowledging which works influence their own. Usage data constitutes a behavioral, implicit measurement of how much "attention" a particular scholarly communication item has garnered. The challenge is thus to turn this type of behavior, implicit, clickstream data into metrics reflecting actual scholarly influence.

(3) Community acceptance: whereas an entire infrastructure is now devoted to the collation, aggregation and disposition of citation data and statistics, usage data remains largely unproven in terms of scholarly impact metrics or services, due to a lack of applications and community services. The challenge here is to create a framework to aggregate, collate, normalize, and process usage data that the community can trust and from which we can derive trusted metrics and indicators.

Enter the MEtrics from Scholarly Usage of Resources (MESUR) project! The MESUR project was funded by the Andrew W. Mellon Foundation in 2006 to study science itself from large-scale usage data. The project was involved with large-scale usage data acquisition, deriving structural models of scholarly influence from said usage data, and surveying a range of impact metrics from the usage and citation it collected.

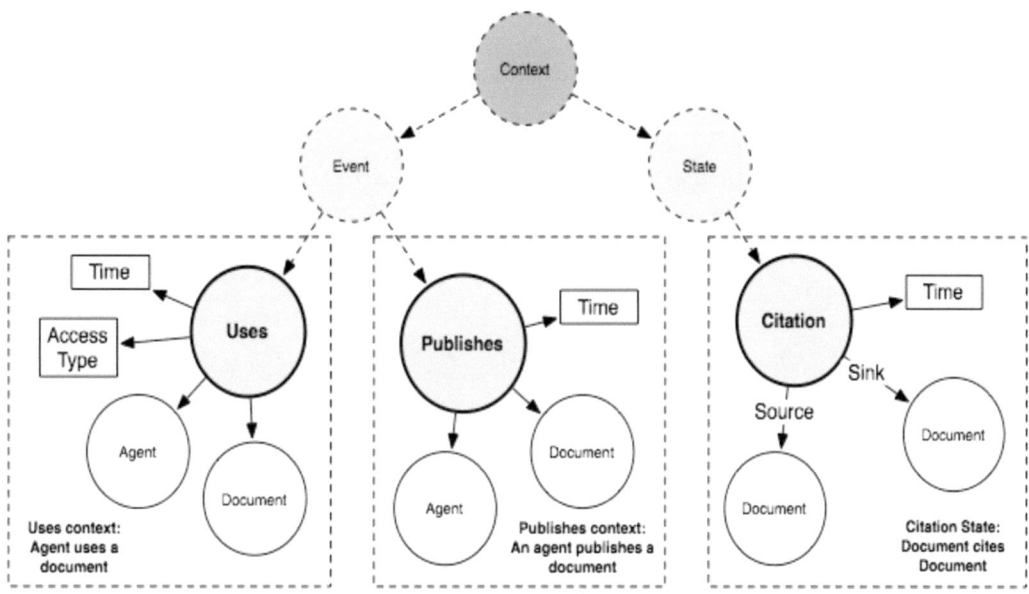

FIGURE 3-2 Modeling the scholarly communication process — the MESUR ontology.[2]

So far, the MESUR project has collected more than one billion usage events[3] from publishers, aggregators and institutions serving the scientific community. These include: BioMedCentral, Blackwell, the University of California, EBSCO Publishing, Elsevier, Emerald, Ingenta, J-STOR, the Los Alamos National Laboratory, Zetoc project of the University of Manchester, Thomson, the University of Pennsylvania, and the University of Texas.

This data provided to the project has to conform to specific requirements, which were fortunately met by all our providers. In particular, we required that the data had an anonymous but unique user identifier, unique document identifiers, data and time of the user request to the second, an indicator of the type of request, and a session identifier, generated by the provider's server, which indicates whether the same user accesses other documents within the same session.

The latter is an important element of the MESUR approach. We are not just interested in total downloads, but their context, the structural features of how people access scholarly communication items over time. We therefore required session identifiers, meaning that if users access a document at a particular time, they are assigned a session identifier before they move on to the next document. They maintain this session identifier throughout their movement from one

---

[2] Marko A Rodriguez, Johan Bollen and Herbert Van de Sompel. A Practical Ontology for the Large-Scale Modeling of Scholarly Artifacts and their Usage, In Proceedings of the Joint Conference on Digital Libraries 2007, Vancouver, June 2007.

[3] Data from more than 110,000 journals, newspapers and magazines, along with publisher-provided usage reports covering more than 2,000 institutions, is being ingested and normalized in MESUR's databases, resulting in large-scale, longitudinal maps of the scholarly community and a survey of more than 40 different metrics of scholarly impact.

document to the next. As a result we can reconstruct so-called clickstreams and model how people move from one document to the next in any particular session. Because we have that kind of data, we can track how users collectively move from one article or journal to the next, and map the collective flow of "scientific traffic." Such a map is shown in Figure 3-3 and was published in PLoS ONE in 2009.[4]

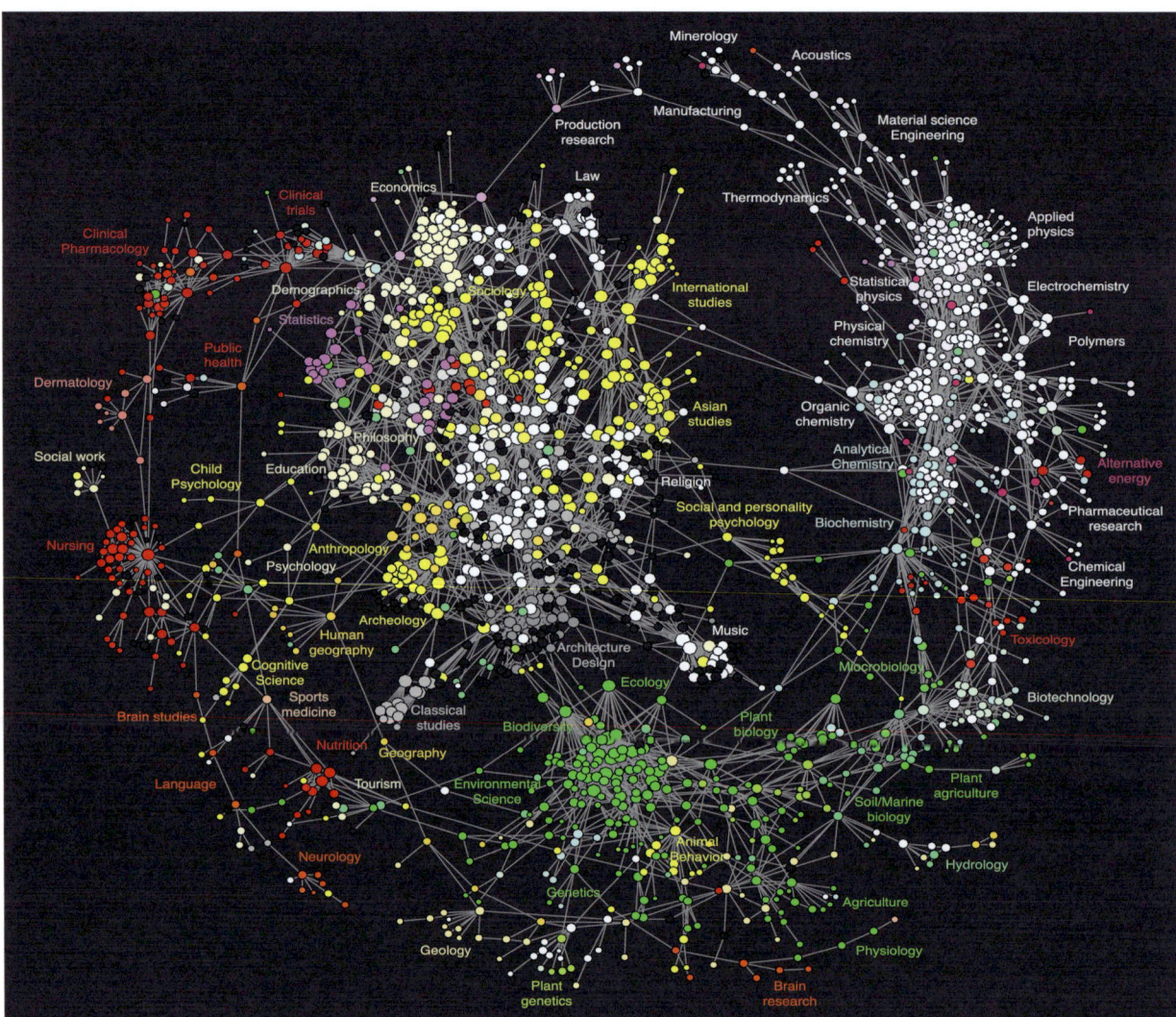

FIGURE 3-3 Visualization of MESUR clickstream data showing how users move from one journal to the next in their online access behavior. Each circle represents a journal. Journals are connected by a line if they frequently co-occur in user clickstreams.

Looking at the map we can see a rich tapestry of scholarly connections woven by the collective actions of users who express their interests in the sequence by which they move from one article and journal to the next in their online explorations. Although from our data we cannot prove that any individual user actually followed a certain path, we can say that it reflects the fact that users

---

[4] Bollen J, Van de Sompel H, Hagberg A, Bettencourt L, Chute R, et al. (2009) Clickstream Data Yields High-Resolution Maps of Science. PLoS ONE 4(3): e4803. doi:10.1371/journal.pone.0004803.

collectively felt these journals are related somehow, leading to the formation of clusters of interests which do not always coincide with traditional domain classifications, cf., the position of psychology journals in this map.

Once we have derived a network structure of related journals from usage data, as shown in Figure 3-3 we can use it to perform the same kind of scholarly assessment analysis that is now commonly conducted on the basis of citation data, and the resulting citation networks. We can actually calculate how important a journal is to the structure of the network, and use it as a measure of scholarly influence or impact.

This is what the MESUR project has done. We surveyed nearly forty different impact metrics, most based on social network analysis. We calculated one half of the metrics from our usage network, and the other half from a citation network that we derived from the Journal Citation Reports. Most usage-based network metrics had a citation-based counterpart. We also added several existing citation-based metrics that are not necessarily based on a citation network, such as the journal's h-index and its Impact Factor. Each of these metrics, depending on whether they were based on usage data or citation data, and their method of calculation, will reflect a different perspective of scholarly impact in the journal rankings it produces. For example, some metrics will indicate how centrally located a journal is in the usage network and serve as an indication of its general impact according to patterns of journal usage. We can also calculate a journal's "betweenness centrality," i.e., how often users or citations pass through the particular journal on their way to another journal from another one. This may be construed as an indication of the journal's interdisciplinary nature, its ability to bridge different areas and domains of interest in the usage and citation network vs. how popular or well connected it is in general. Each metric by virtue of its definition will have something different to say about a journal's scholarly impact, and can furthermore be calculated from either usage networks or citation networks, offering even more perspectives on the complex notion of scholarly impact. A comparison of all of these metrics was published in PLoS ONE in 2009, and yielded a model of the main dimensions along which scholarly impact can fluctuate.[5]

We are also working on a number of online services to make our results accessible to the public. As mentioned, the problem with this kind of usage data is that people have a hard time accepting its validity since citation data is so ingrained. Usually, I get arguments such as "You may have nice results, but I don't believe it." A public, open, freely available service will allow people to play with the data and results themselves and might make them more community accepted.

Finally, I want to mention that we secured new funding in 2010 from the Andrew W. Mellon Foundation to develop a generalized and sustainable framework for a public, open, scholarly assessment service based on aggregated large-scale usage data, which will support the evolution of the MESUR project to a community-supported, sustainable scholarly assessment framework. This new phase of the project will focus on four areas in developing the sustainability model:

---

[5] Bollen J, Van de Sompel H, Hagberg A, Chute R (2009) A Principal Component Analysis of 39 Scientific Impact Measures. PLoS ONE4(6): e6022. doi:10.1371/journal.pone.0006022.

financial sustainability, legal frameworks for protecting data privacy, technical infrastructure and data exchange, and scholarly impact. It will integrate these four areas to provide the MESUR project with a framework upon which to build a sustainable structure for deriving valid metrics for assessing scholarly impact based on usage data. Simultaneously, MESUR's ongoing operations will be continued with the grant funding and expanded to ingest additional data and update its present set of scholarly impact indicators.

I would like to end my presentation by highlighting the following interesting initiatives and some relevant publications.

Initiatives:

- Microsoft/MSR: http://academic.research.microsoft.com/
- Altmetrics:http://altmetrics.org/
- Mendeley-based analytics: using Mendeley's bookmarking and reading data to rank articles.
- Publisher-driven initiatives: Elsevier's SciVal , mapping of science: http://www.elsevier.com/wps/find/authored_newsitem.cws_home/companynews05_01743
- Google Scholar : http://scholar.google.com/
- Science of Science Cyberinfrastructure: http://sci.slis.indiana.edu/ (Katy Borner at Indiana University)

Relevant Publications by the Presenter:

- Johan Bollen, Herbert Van de Sompel, Aric Hagberg, Luis Bettencourt, Ryan Chute, Marko A. Rodriguez, Lyudmila Balakireva. **Clickstream data yields high-resolution maps of science.** PLoS One, March 2009.
- Johan Bollen, Herbert Van de Sompel, Aric HagBerg, Ryan Chute. **A principal component analysis of 39 scientific impact measures.** arXiv.org/abs/0902.2183
- Johan Bollen, Marko A. Rodriguez, and Herbert Van de Sompel. **Journal status.** Scientometrics, 69(3), December 2006 (arxiv.org:cs.DL/0601030)
- Johan Bollen, Herbert Van de Sompel, and Marko A. Rodriguez. **Towards usage-based impact metrics: first results from the MESUR project.** In Proceedings of the Joint Conference on Digital Libraries, Pittsburgh, June 2008.
- Marko A. Rodriguez, Johan Bollen and Herbert Van de Sompel. **A Practical Ontology for the Large-Scale Modeling of Scholarly Artifacts and their Usage,** In Proceedings of the Joint Conference on Digital Libraries, Vancouver, June 2007.
- Johan Bollen and Herbert Van de Sompel. **Usage Impact Factor: the effects of sample characteristics on usage-based impact metrics.** (cs.DL/0610154)
- Johan Bollen and Herbert Van de Sompel. **An architecture for the aggregation and analysis of scholarly usage data.** In Joint Conference on Digital Libraries (JCDL2006), pp. 298-307, June 2006.
- Johan Bollen and Herbert Van de Sompel. **Mapping the structure of science through usage.** Scientometrics, 69(2), 2006.

- Johan Bollen, Herbert Van de Sompel, Joan Smith, and Rick Luce. **Toward alternative metrics of journal impact: a comparison of download and citation data.** Information Processing and Management, 41(6): 1419-1440, 2005.

4- Data Citation —Technical Issues— Identification

Herbert Van de Sompel[1]
Los Alamos National Laboratory

I am going to speak today about a slightly narrow topic, which is about the identification of data as part of the citation process. I will also talk about different use cases, such as assigning credit, accessing and reusing the data, and involving both humans and machines. To make clear what I am talking about, I will give some alternate examples of a citation:

- *Bollen, J., Van de Sompel, H., Hagberg, A., Chute, R. A Principal Component Analysis of 39 Scientific Impact Measures. PLoS ONE, 4(6), pp. e6022, 2009.*
  doi:10.1371/journal.pone/0006022
- *Bollen, J., Van de Sompel, H., Hagberg, A., Chute, R. A Principal Component Analysis of 39 Scientific Impact Measures. PLoS ONE, 4(6), pp. e6022, 2009.*
  doi:10.1371/journal.pone/0006022 http://dx.doi.org/10.1371/journal.pone/0006022
- *Bollen, J., Van de Sompel, H., Hagberg, A., Chute, R. A Principal Component Analysis of 39 Scientific Impact Measures. PLoS ONE, 4(6), pp. e6022, 2009.*
  http://dx.doi.org/10.1371/ journal.pone/0006022

These are the three considerations that merit more attention:

- The nature of identifiers for citation, access, and re-use.
- Catering to human and machine agents.
- Granularity, both spatial and temporal.

First, let us examine the nature of identifiers for these different use cases:

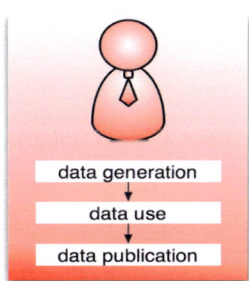 

FIGURE 4-1 Data publication—data citation

---

[1] Presentation slides are available at http://sites.nationalacademies.org/PGA/brdi/PGA_064019.

At the left-hand side, we have someone who is sharing data (e.g., data generation, data use, and data publication) and, at the right-hand side, there is someone who wants to do something with the data (e.g., access the data, reuse the data, and cite the data). The first consideration is about the nature of the identifier. We have these two parties and in order for the consumer to access, use and cite the data, some information needs to be made available by provider. I am going to distinguish between identifiers that enable citation and identifiers that enable access and reuse.

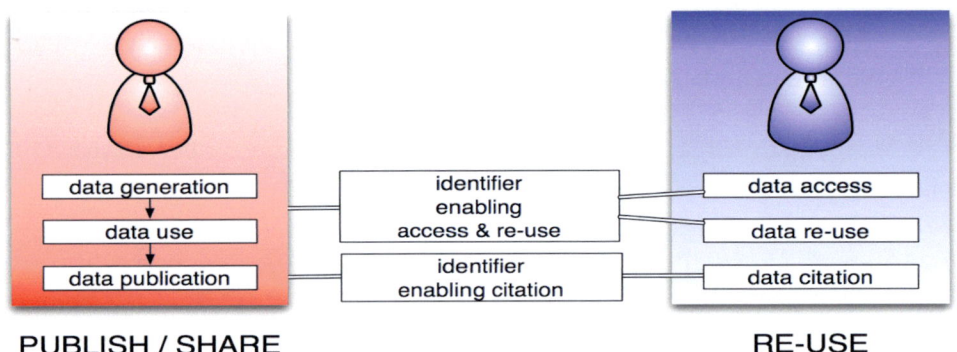

FIGURE 4-2 Identifiers that enable citation, access, and reuse.

If you look at the identifiers that enable citation, we can think of two choices. We have identifiers like the DOIs that are used extensively, but we also have the choice for a HTTP URI. Moreover, an HTTP URI could be based on a DOI (i.e., the HTTP version of a DOI) or it could be any other stable, cool HTTP URI.

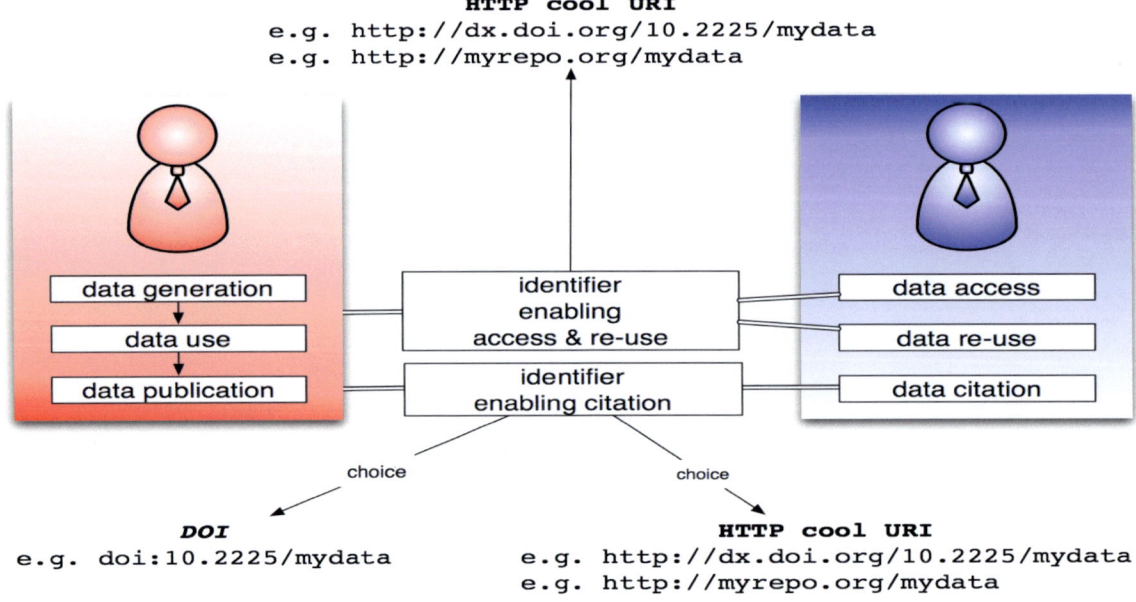

FIGURE 4-3 Actionable and non-actionable identifiers for citation, access, reuse.

When thinking about identifiers for enabling accessing and reuse, there is only one choice because the focus with this regard is on access. Access means an identifier that is actionable using widely deployed technologies such as web browsers, web crawlers, etc. This yields the use of HTTP URIs for access. So, DOIs as such can be used for citation, whereas cool HTTP URIs (including the HTTP version of a DOI) can be used for citation, access, and reuse. This was the first consideration.

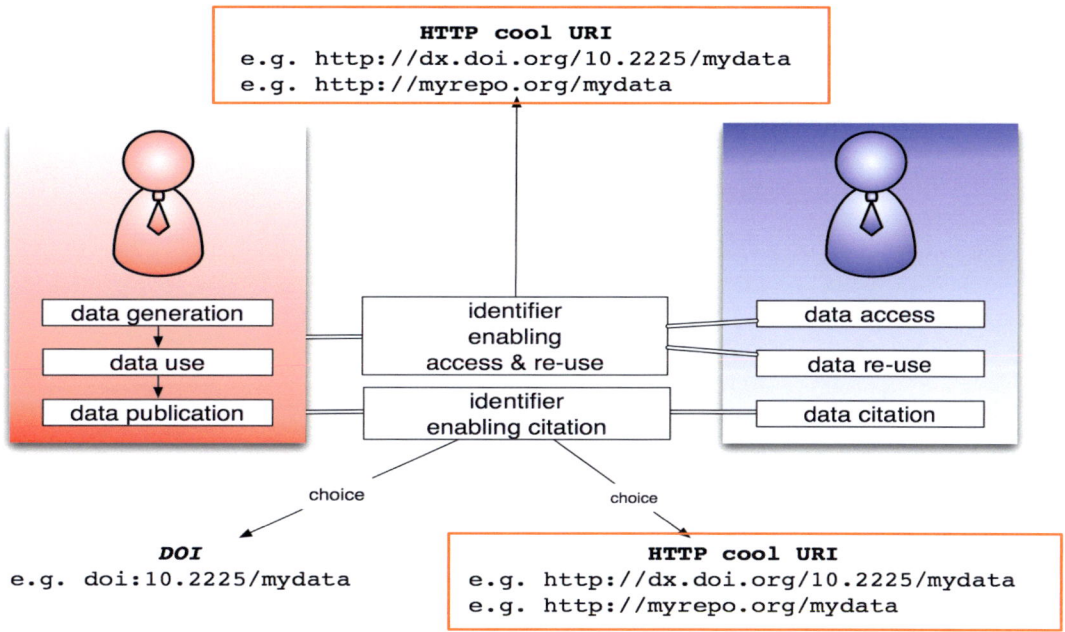

FIGURE 4-4 Cool HTTP URIs can be used for citation, access, and reuse.

The second consideration is about catering to human and machine agents when talking about accessing and reusing data. When it comes to papers, the typical end-user is a person. However, when it comes to data, the most important consumer most likely is going to be a machine. It is not clear at this point how machine agents will be enabled to access and reuse the data. This means that we need to think carefully about aspects such as links, metadata, and discovery measures to cater to machines.

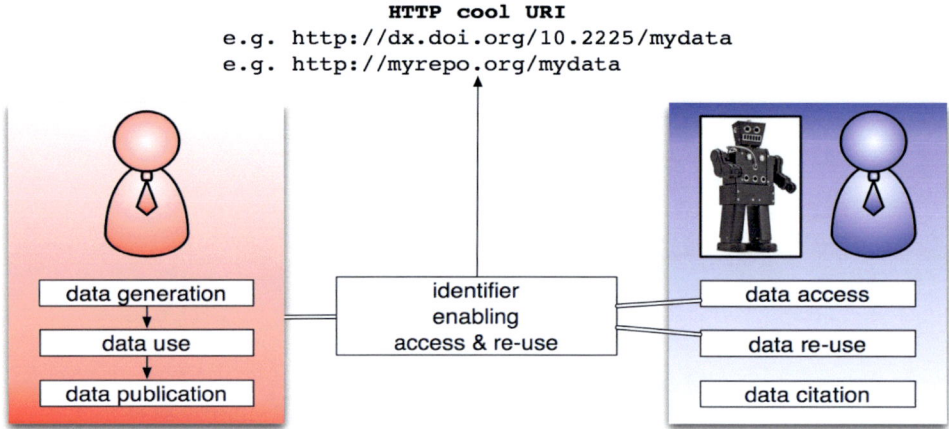

FIGURE 4-5 Data access and reuse by both human and machine agents.

As you know it today, we have a URI that sits somewhere in a citation and these are the reference data sheets for humans and machines.

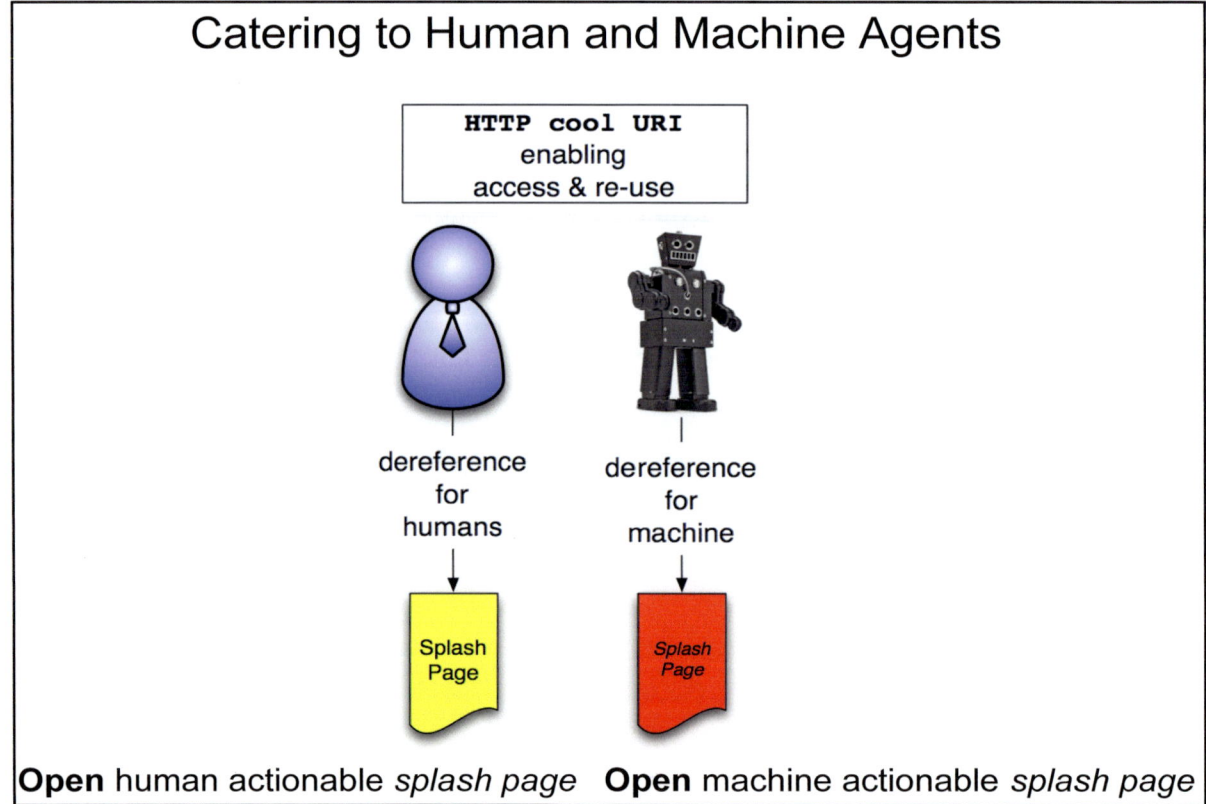

FIGURE 4-6 Splash pages for humans and machines.

When it comes to machine processing, there is some critical thinking that needs to be done about what has to be involved there. This is not only about discovery of the data or metadata that supports discovery of the data. This is about metadata that supports understanding of the data and interacting with the data in automated ways.

For example, for any dataset, there can be several URIs involved, including URIs for the splash pages (one for human and one for machine agents) and URIs for multiple components of the dataset. Maybe there are different formats of the same data available, in which case there are even more URIs and the need to express a format relationship. We have to distinguish between all those URIs and make sure that it is clear to the machine what each URI stands for. We need technical metadata about the data so that the machine can automatically interact with the data, process it. Also, not all data is necessarily being downloaded; some is accessed via APIs. In this case, a description of those APIs is required. Overall, a rich description of the data that is able to be processed by machines is required.

The third consideration is granularity and I am going to distinguish between two types. One is spatial and by spatial I mean dimensions in the dataset, and the other is temporal.

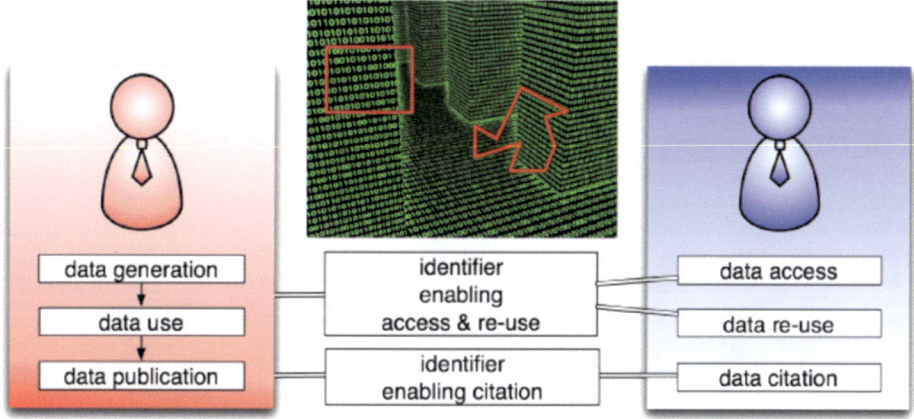

- Simple segments (dimensions in hypercube)
- Complex segments (*arbitrary* queries)

FIGURE 4-7 Identifying, accessing, and reusing parts of a dataset.

Regarding spatial granularity: There is a need to say I have used this entire data set, but I actually only worked with the parts of it. It is important to be able to describe that slice or segment of the dataset.

I would like to draw some parallels here with ongoing work related to web-centric annotation of research materials as pursued by the Open Annotation and Annotation Ontology efforts. What you see in Figure 4-8 is an image that is being annotated. It is not the entire image that is annotated; it is just a segment thereof. And that segment is being described, in this case by means of an SVG document. A similar approach could be used when referring to parts of a dataset:

identify the dataset, and convey the part of the dataset that is of interest by means of an annotation to the dataset.

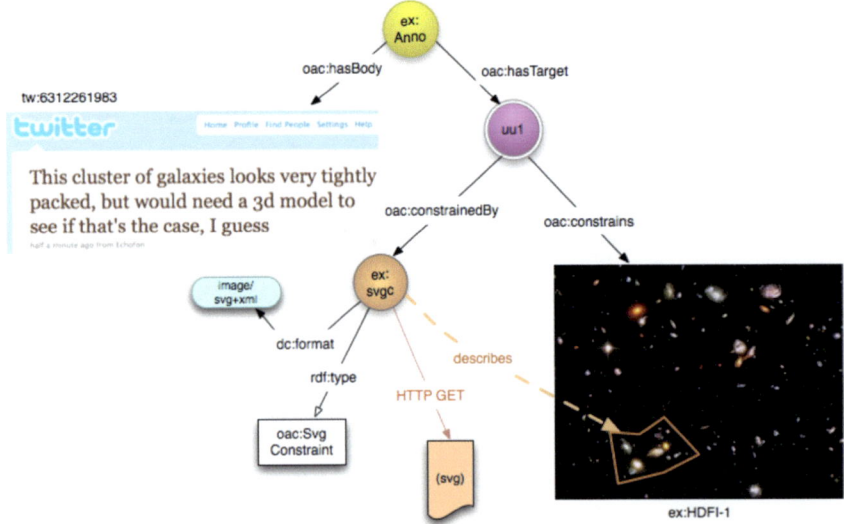

FIGURE 4-8 An annotation on a part of an image per the Open Annotation approach.

Regarding temporal granularity: We need to be able to refer to a certain version of a dataset as it changes over time, or a certain temporal state of the dataset when a dynamic dataset is concerned.

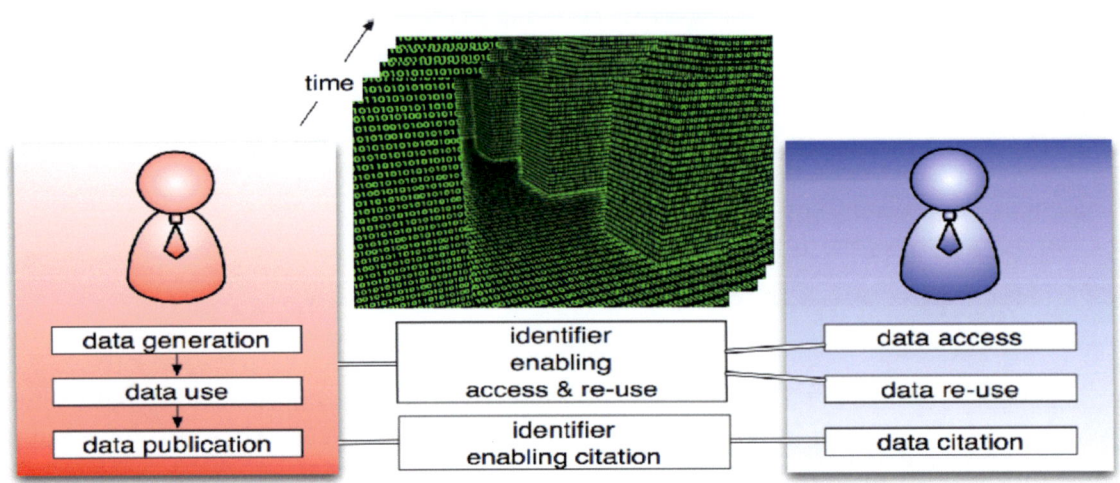

- Different versions of dataset over time
- Different states of dataset over time (dynamic data)

FIGURE 4-9 Dataset changes over time.

Here, again, I would like to draw a parallel with other work that I am doing. This is the Memento project, which is a simple extension of the HTTP protocol that allows accessing prior versions of resources. In order to do this, Memento uses the existing content negotiation capability of HTTP and extends it into the time dimension. So, what you have in Figure 4-10 a generic URI for a certain resource (URI-R). But the resource evolves over time and each version receives its own URI (URI-M1, URI-M2). Without Memento, you need to explicitly know URI-M1 and URI-M2 in order to access those respective versions. With Memento, you can access the old versions of the resource, only knowing the generic URI (URI-R) and a date-time. Memento can play a role in accessing prior versions of datasets.

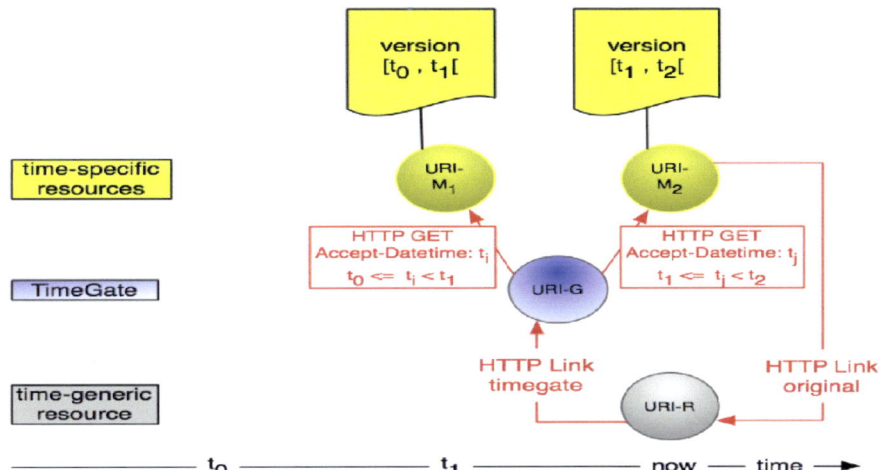

- Similarity with Memento protocol that allows accessing prior versions/states of a resource given its original URI

FIGURE 4-10 Memento allows to access old versions of resources using a generic URI and a date-time.

I would say that these granularity requirements are needed for both the data publisher and the data user. If we want to address those granularities, we need a solution that allows both the publisher and the user of the data to specify what that segment is going to be.

Finally, I see three options regarding identity in regarding the notion of granularity:

- Option 1. For citation, access and reuse: Mint a new identifier for each segment.
- Option 2. For citation, access and reuse: Use the entire dataset identifier with a query component.
- Option 3.
    - For citation: Use the entire dataset identifier.
    - For access and reuse: Use an additional URI that refers to an annotation to the dataset that describes the segment (both spatial and temporal) of the dataset. The citation thus becomes a tuple [dataset URI; annotation URI].

5- Maintaining the Scholarly Value Chain: Authenticity, Provenance, and Trust

Paul Groth[1]
VU University of Amsterdam, The Netherlands

I think the following quote is important. It is from Jeff Jarvis[2]:

> In content, as creation becomes overabundant and as value shifts from creator to curator, it becomes all the more vital to properly cite and link to sources [...]. Good curation demands good provenance. [...] Provenance is no longer merely the nicety of artists, academics, and wine makers. It is an ethic we expect".

I agree that this is an ethic that we expect and that is one of my motivations for doing research on provenance.

I am a computer scientist. One of the design principles that computer scientists adopt a separation of concerns. I think that when we talk about data citation, we need to be very careful about separating concerns. This lack of separation of concerns occurs because when we speak about data citation we adopt practices from the way we cite traditionally. My goal is to convince you that we can do better.

Let me give you an example. The figure below is a standard reference from an older famous paper in the field of artificial intelligence titled "A Truth Maintenance System".[3]

FIGURE 5-1 A reference from the article, "A Truth Maintenance System."

This reference contains a lot of information. It tells you where to find this article. It gives you a search term to help you find it in libraries or in Google. It also helps with provenance. For example, it gives information about the person who wrote this paper and that it appeared in a book called the Fourth Proceedings of IJCAI, as identified in Figure 5-2.

---

[1] Presentation slides are available at http://sites.nationalacademies.org/PGA/brdi/PGA_064019
[2] Jeff Jarvis, media company consultant and associate professor at the City University of New York's Graduate School of Journalism, in "The importance of provenance", on his BuzzMachine blog, June, 2010.
[3] Jon Doyle, A truth maintenance system, Artificial Intelligence, Volume 12, Issue 3, November 1979, Pages 231-272, ISSN 0004-3702, 10.1016/0004-3702(79)90008-0.

FIGURE 5-2 Provenance information with a reference.

Furthermore, the reference helps with making trust interpretations. For example, I personally know that Pat, the author, is a good person. This means that I should follow this citation because I know this person is trustworthy. Or maybe, I should follow the citation because it is at the International Joint Conference on Artificial Intelligence and I happen to have background knowledge about artificial intelligence to know that this is a top conference in the field. So, again, I think I trust this piece and think it can be used in my work. The point is that in this one simple citation and its corresponding reference, we were provided with information about four different things: information to lookup the paper, its identity, provenance information, and trust. However, giving this same reference to a computer changes things completely.

I think that for data citation we need to try to address all of these different areas separately. We need identity and we need provenance, but provenance is not part of making persistent identifiers. It is something separate. Once we have provenance, we can start computing trust metrics. Different people have different ideas about what they trust, but based on where the material comes from and describing how the experiments were produced we can compute different kinds of trust metrics. Finally, over the top of all this, once we have identifiers, once we have some information about where the material comes from and how it was produced, once we have these trust metrics, we can build good search engines.

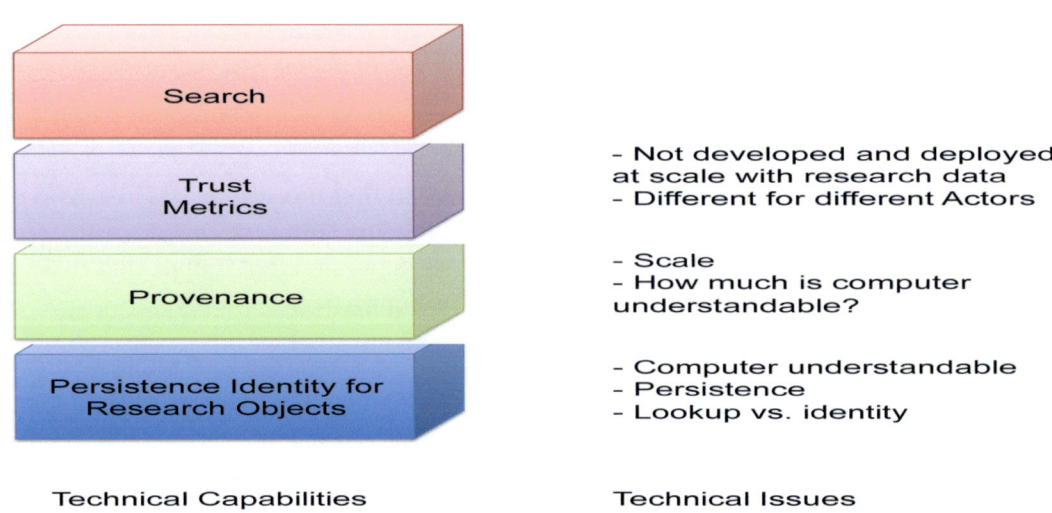

FIGURE 5-3 Technical capabilities vs. technical issues.

Each area is a different concern and there are some technical issues with each one. For identity, we have already heard about computer understandability, about identity persistence and also the notion of lookup versus identity. Whether we need to put together or pull apart, lookup and identity, I do not know.

For provenance, we need to deal with the issue of scale. One data point can have hundreds of gigabytes of provenance. I remember when I did my Ph.D., I was doing provenance work and I had a result that was one number with one gigabyte of provenance information associated with it. So scale is a huge issue here, especially how much of this is computer understandable? Right now most of what we say about provenance of data is encoded in text that people can read, but increasingly we encode in the form of computational workflows or in other computer reasonable formats.

It is important to note, that trust is different than provenance. Provenance can be viewed as a platform on which we can develop different ways of determining trust. I think we need to develop this platform trust metrics separately. We also need to develop scalable algorithms for calculating these trust metrics. Finally, I think trust is different for different actors and, right now, we do not have that clear a conception of it. So this also requires more work. For example, for my many data applications, I may trust everything because statistically it does not matter. However, for something that my tenure or promotion package is based on, maybe I want a different, higher level of trust.

To conclude, this is my appeal: Citation does not have to contain everything. When we talk about data citation, we should not try to include everything possible in the citation or reference. Maybe all we need is simple pointers that are understandable by machines. The final thought I want to leave you with is that we can build it. We have the technology and I do not think we are far off. I will end with some references:

W3C Provenance Incubator Final Report:
http://www.w3.org/2005/Incubator/prov/XGR-prov-20101214/.

W3C Provenance Working Group Standardization Activity:
http://www.w3.org/2011/prov/wiki/Main_Page.

Surveys of provenance and trust:

Donovan Artz and Yolanda Gil. A Survey of Trust in Computer Science and the Semantic Web, Journal of Web Semantics, Volume 5, Issue 2, 2007.

Rajendra Bose and James Frew. Lineage Retrieval for Scientific Data Processing: A Survey. ACM Computing Surveys, Volume 37, Issue 1, 2005).

J. Cheney, L. Chiticariu and W.-C. Tan. Provenance in databases: Why, where and how, Foundations and Trends in Databases, 1(4):379-474, 2009.

David DeRoure, Replacing the Paper: The Twelve Rs of the e-Research Record: http://blogs.nature.com/eresearch/2010/11/27/replacing-the-paper-the-twelve-rs-of-the-e-research-record

Juliana Freire, David Koop, Emanuele Santos, Claudio Silva. Provenance for Computational Tasks: A Survey, Computing Science and Engineering, Vol 10, No 3, pp 11-21, 2008.

Luc Moreau, The Foundations for Provenance on the Web, 2010, Foundations and Trends® in Web Science: Vol. 2: No 2-3, pp 99-241. http://dx.doi.org/10.1561/1800000010

Yogesh L. Simmhan, Beth Plale, Dennis Gannon. A survey of data provenance in e-science. ACM SIGMOD Vol 34, No 3, 2005. See also a longer version.

DISCUSSION BY WORKSHOP PARTICIPANTS

Moderated by John Wilbanks

PARTICIPANT: I did not get the distinction between usage data, citation data, and bibliography data. Can you please define these terms?

DR. BOLLEN: The least we can say about usage data is that someone paid attention to a particular resource, whereas with citations, it is an explicit public statement of someone indicating that they have been influenced or impacted by someone else's work. In our statistics, it is sometimes obvious that some materials are used a lot but never cited and vice versa. When we raise this data usage issue, we also talk about data downloads and access, and we are focusing on aspects that are somehow measurable.

PARTICIPANT: I believe I got your distinction between usage data and citation data, but I think that this is not well developed. We might need to make a distinction between citation and usage of data versus citation and usage of literature.

PARTICIPANT: I think that the issue of data citation has two components: technical and socio-cultural. We still have questions related to how we write a citation for a dataset, and there are some technical challenges in terms of space or in terms of the style of writing the citation for a given dataset. However, are we focusing too much only on the technical challenges? The real challenge is the culture of citing datasets and providing a proper citation. We therefore need to focus on both the technical and the social and cultural challenges.

DR. MINSTER: There is a well-known trick to put the subtle error in one of your papers and then everybody is going to cite your publication. You can do the same with data. If we stop insisting that every single granule of a database should have its own identifier and be quoted by those who use it, it would be in the interest of the data provider to just give miniscule granules.

PARTICIPANT: Why can I not point to every piece of data at the smallest granularity possible?

DR. BOLLEN: Fair enough, but if you rely on this mechanism to provide people with recognition, advancement, tenure, things like this, the small cites are not very useful.

PARTICIPANT: This question is for Dr. Bollen. Did you put your usage data in your tenure package?

DR. BOLLEN: Yes, I did. I had a couple of pages showing some statistics on my papers.

PARTICIPANT: Did you make the raw data available?

DR. BOLLEN: No. I could not do that because of the agreement that we have to sign with the rights holder. This issue is always a big challenge and a big responsibility.

PARTICIPANT: One of the great things about print citations is that they are notation dependent. In the past, if you happened to work at a university, you knew how to go to the library and look up a journal title and get it. Those practices survived hundreds of years, and I worry sometimes that if we revert to just URIs and require that it should be de-referenced, we are losing something very important from the old model. Do you have any thoughts about this tradeoff between location and dependence on simplicity?

DR. BOLLEN: This is a good question. The best answer I heard this morning was that when you want to deal with regularity, you have to retrieve the same dataset that somebody else used. Maybe storing separately the data and the dataset of the query is the right way to do it. Then you can say, "in this research, I have used the output of this query on that dataset".

PARTICIPANT: Where would I find the dataset? Where do I look?

DR. BOLLEN: The dataset should be findable by URI on the web, no matter where it is, although some providers may require enormous granules or miniscule granules. Having that as the output of the query would be a reasonable thing to do.

DR. GROTH: I think this is one of the places where we may have to sacrifice because of the scale we are talking about. The reason we can look at a normal citation and find it in the library is because there is lots of background information about what all the pieces of a citation mean. In the data area, it is not that clear. We had a great example of what a possible data citation could look like in a paper, but we already have been told that it is too much work. If we think about the scale we are talking about, then URLs may even be too much work. So we have to sacrifice this location independence.

DR. VAN DE SOMPEL: I am in favor of the HTTP URIs in this case. I have some reasons for that. One, we get an entire infrastructure that is freely available. We can get all the new developments free rather than having to reinvent the wheel. In order to get closer to what we refer to as longevity of identifiers similar to what we have in print, I am actually in favor of having an accession number that is not protocol based, like the DOI or an ARK identifier as introduced by the California Digital Library. It is just a unique string that you carry with you into the future. I call this technology independent. You can print it on paper. Then you basically instantiate that non-protocol identifier and use the protocol of the moment. So, even if all those citations become invalid at one point because HTTP goes away, there will be ways to recover that information and transpose it to the next protocol. Plus, if indeed HTTP is going to become obsolete at some point, there are going to be services to migrate us from one to the other. We just have to rely on that. So, yes I am now in favor of using HTTP URIs, but probably under the condition that the non-protocol string identifier exists also.

DR. BOLLEN: I share that concern. There is definitely going to be some problem in this area. I am not a specialist on identifiers, but a lot of this seems to me to be an artifact of the technology that we have at our disposal right now. If we had machine intelligence that could, in essence, unpack the kind of information that we use, none of this would be an issue.

PARTICIPANT: However, humans do create identifiers.

DR. BOLLEN: This involves a lot of background knowledge and implicit information. The reason why we are coming up with these machine identifiers specifically, however, is because the machines are not capable of accessing that information.

DR. SPERBERG-MCQUEEN: We should not overestimate the power of human contextual information. Bear in mind that in difficult cases, in particular in identifying musical compositions, we always use catalog notes. You do not refer to a piece by Mozart without a peripheral number precisely because context does not suffice for average use in those situations.

PARTICIPANT: Dr. Minster, you referred to something being "properly cited." What did you mean by that?

DR. MINSTER: I used "properly cited" in a loose way, but I believe it has two purposes. The first purpose is to give appropriate credit to the person who actually created the data, analyzed them, calibrated them, or did whatever was necessary before they become public. The second is for someone else to be able reuse these data and do his/her own analysis.

PARTICIPANT: I have another question for Dr. Minster. What did you mean by "data publication?" Are you talking about fixing data in time or printing it on paper? Or are you talking about just making it available in some other form? Or are you actually talking about a publication like a data journal?

DR. MINSTER: No, I do not mean a data journal, but I do mean something that may have a versioning capacity. For instance, if we want to refer to a book, we have to say which edition we want. That is the same thing about data. There is another aspect, however, which I am not sure how to solve. In our world today, we trust the gold standard for scientific publication: peer review. I do not know how one would do a dataset peer review. This is something that might have to be invented in the future.

PARTICIPANT: Does citation have anything to do with peer review? That is the thing that I do not really get. Just being able to point to something formally and give it credit should be fine. We need to be able to do that first and then we may need new mechanisms for peer review of data.

DR. MINSTER: The question was about publication, however. I have a hard time accepting the concept of citable publication without some kind of quality control, either peer review or an equivalent mechanism.

PARTICIPANT: When a database comes out, it is often transformed before it is published. That transformation can then be done later by a third party and, in fact, there could be a whole new data chain. At what level of granularity does it make sense to track this into a machine-understandable form so that we can say, "I am looking at this transformation of this data by this person?"

DR. GROTH: I have worked with people who track the provenance of data at the operating system level. I think it is the decision of scientists. There is no hard and fast rule for that.

DR. MINSTER: In my field of geophysics, more and more datasets are the output of a large computer program and, in that case, provenance is a real issue because we have to say which version of the code we used, on what machine, and using what operating system.

DR. VAN DE SOMPEL: I agree with Paul Groth that this is a more curatorial decision. I am not an expert in this area, but it seems to me that we can hold on to all the provenance or workflows that have taken place. The question, though, is do we also need to hold on to all intermediate versions of the data? Are these workflows still relevant when we do not have the underlying data anymore? Maybe they are because we can figure out what had happened even though the source data are gone.

DR. GROTH: In general, if we have un-deterministic data, we can throw away the intermediate data. But if we have un-deterministic data or information, we cannot throw away steps because we cannot reproduce them. We did some work in astronomy, where we put everything on a virtual machine and anyone could just re-execute everything. It was fine because we had all the input data and everything could be reproduced.

DR. CALLAGHAN: I want to respond to Paul Groth's comment on the relationship between data publication and data citation. I think that we need to have citation before we can have data publication. This is because the way we are looking at publishing data (i.e., actual formal journal publishing with the associated peer review arrangements) is to give it that cited stamp of quality. We cannot do that unless the dataset that is being peer reviewed is findable, static, a host of other things, and is persistent.

I work for a data center and we do what some people might consider data publishing. It gets a bit confusing because we have two levels of publishing. There is publishing that involves making labels, handing it out to the web, and the like. We call it publishing with a small "p". Then there is publishing that carries all the connotations of scientific peer review, quality, persistence, and so on. We call that publishing with a big "P." I view citation of data as occurring somewhere between those two. So, from my point of view, citation means that as data centers, we are making some judgments about the technical quality of the datasets. We cannot say anything about the scientific quality of the data because that is for the domain experts and the scientific peer review process to decide.

DR. GROTH: When we use a citation in a journal article, I know that in geography, for example, they only cite things that are peer reviewed. In computer science, however, if you examine the articles produced by the logic community of practice, they cite their own technical reports because their proofs are so long that they cannot put them in the paper itself. So, I think we need to allow for citation of the small "p" and that should be the same mechanism for the citation of the big "P."

DR. CALLAGHAN: I have a quick point related to the granularity issue and citing a small part of the dataset. The way I think about it is that we do not need to have a DOI or a URI for every word in a book, for example. We cite the books and some location information at the end, such as page number, paragraph number, or something similar. That is the analogy I would think of when it comes to citing a portion of the dataset. We cite a dataset as a whole, but we can also provide extra information to locate the specific part of the data that we want to reference.

DR. VAN DE SOMPEL: That is exactly the formulation that I showed on my presentation slides, which I feel is also the most generic approach to this: use the identifier of the entire thing and in addition use an annotation (with an identifier) to express which part of the thing is of concern. All of this is easier when using HTTP URIs, including ones that carry DOIs or ARK identifiers, because a whole range of capabilities comes freely with HTTP.

MS. CARROLL: We got separated from the publication and citation issues, and I would like to get back to them. I think we have to have a common understanding of what we are talking about when it comes to publication. I think the division between scholarly articles and technical reports is a good analogy. The vast majority of datasets that people want are like technical reports rather than like scholarly journal articles. So, let us not lose sight of those little "p's".

DR. VAN DE SOMPEL: I agree and I would like to reiterate that peer reviewing in the classic way is not going to be scaleable in the area of small "p."

MR. WILBANKS: I think it is also important to note that even in the big "P," traditional peer review is starting to change. Recently, we started to see articles getting published that are only reviewed for scientific validity, without attempting to judge input in advance. This is not just in the library science area; it is in *Nature* and is used by some other publishers.

DR. CALLAGHAN: I think that what is really important is how people actually use the material, regardless of whether it has gone through a formal review process or not. If we are looking to the future and how people are collaborating and communicating, that whole formal scholarly review and the metrics of it should be reconsidered.

DR. BOLLEN: That is exactly the point. I think that within the next 10 or 15 years, we will see the momentum of building new systems and mechanisms and that the traditional big "P" publication process will be slowly but surely replaced by new approaches.

MR. PARSONS: Continuing on this peer review theme, one of the things that I think is critical in this area is training in relation to the concept of data citation. I have been pushing data citation for more than a decade and only recently, in the last couple of years, has the issue really taken off. The scholarly literary mechanisms will likely transform, but also let us be honest and admit that academia is an incredibly conservative institution. It does not transform quickly and one of the reactions I have gotten in pushing data citation is that the data producers do not want the data to be cited. They want the paper about the data to be cited, because they get more credit for that.

PARTICIPANT: Just a quick clarification, Dr. Bollen. I thought that all of the materials you were studying were big "P" publication.

DR. BOLLEN: Yes, we were heavily focused on cross-validation of our results and the only comparable datasets that we had were citation statistics, which do focus on big "P" publications.

I am not expert on persistent identifiers, but I feel that these issues were working themselves out into the larger community by mechanisms that are very difficult to anticipate at this particular point. I think they will be very different from the publication and the scholarly communication mechanisms that we have seen in the past. This is my personal experience and that is true for my students. That is also true for all of my colleagues. A lot of the material that they publish these days is distributed primarily online, and is reviewed primarily online by the general community and not by an editor or a committee of three or four reviewers. I see this phenomenon coming into play when it comes to datasets that are increasingly made publicly available, but not in any systematic way anytime soon.

DR. GROTH: I have three comments. One, I think that we need a simple and straightforward way to point to small "p" and big "P" datasets. Two, once we have that, we can use it as the basis for building more complicated systems around provenance and later on, trust. So, let us start with the simple task first. Finally, we may need some standard ways to make citations look nice in the back of our papers.

DR. MINSTER: It is a sad reality that some of our colleagues work very hard to produce the datasets and they get no credit. Once we have done a piece of work and published it with a big "P," it ends up in a journal or in a book. How often is it that you get somebody to hack the book and change the contents of, for example, *Science* magazine online? I just do not see this happening. In datasets, however, this could happen all the time and therefore it is very important to have trusted repositories.

PART TWO

DISCIPLINE-SPECIFIC ISSUES

6- Towards Data Attribution and Citation in the Life Sciences

Philip Bourne[1]
University of California at San Diego

My talk today will focus on some observations about data citation and attribution from the life sciences perspective. The Protein Data Bank (PDB) is a data repository that I am involved with and I am going to use it an example of some of the things that are happening with data in the life sciences.

Let me start with the following observation. In terms of life sciences data repositories, the National Library of Medicine (NLM) is one of the largest in the field. In many ways, the NLM has done a very good job of providing data resources. It also allows people to deposit data and has some level of integration with the literature. However, it appears to me that what they are doing is not fully consistent with some of the data citation and attribution principles and best practices that have been discussed in this workshop so far.

For example, the ability to cite and attribute the data at the NLM is highly variable:

- Digital Object Identifiers (DOIs) are assigned in some cases, but are not used.
- Attribution is made through the metadata in most cases.
- Such attribution is typically through the associated literature reference, if it exists, or by a database identifier.
- The use of data repositories such as Dryad is compelling for the "long tail" problem, but not recognized by NLM.
- Data journals are on the horizon.

There is an interplay between data and publication that is very interesting in the life sciences. To people that maintain data repositories, their metric of success is being published in Nucleic Acids Research, since it brings prominence to these data resources through traditional publication. Similarly, I am an author of a paper about the PDB, which is cited yet I can guarantee that few people have ever read it. There is no reason to read it. It is just there to provide a conventional value metric for a database.

The PDB is a resource that is distributing worldwide the equivalent to one quarter of the Library of Congress each month. The PDB is one of the oldest data repository in biology, a 1 TB resource that is used by approximately 280,000 individuals per month, including an increasing number of school kids. There is absolutely no way that when this resource was founded over forty years ago that scientists ever believed that school children would be using that data.

---

[1] Presentation slides are available at http://sites.nationalacademies.org/PGA/brdi/PGA_064019.

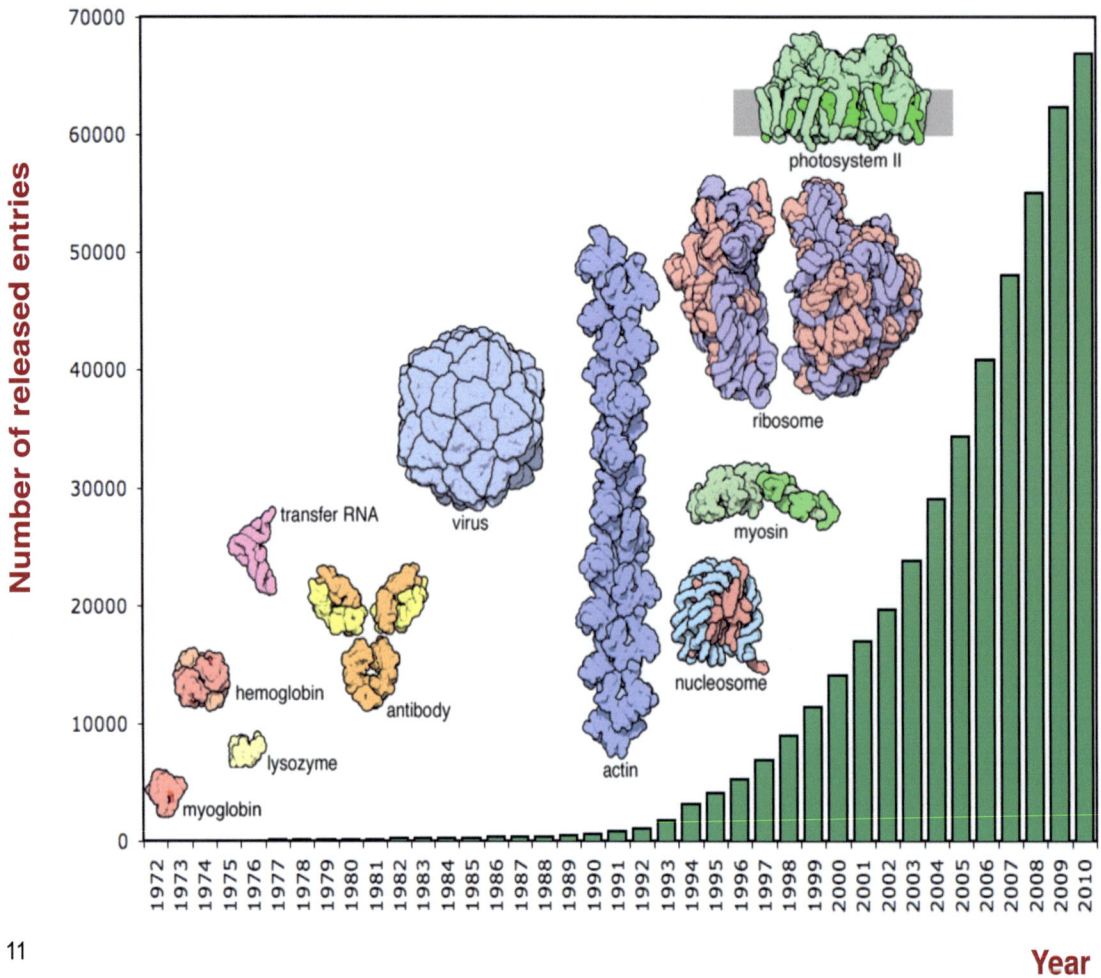

FIGURE 6-1 Protein data bank repository growth over time.

The main point in this diagram is that PDB data volumes are increasing and that the complexity of the data is increasing dramatically as well. Therefore, how we define data, how we cite them, and what granularity we use is changing all the time.

Another point I would like to make is that people are doing more with the data. The following is our big metric of success that we use when we go to funding agencies. We have grown the user base considerably and people are spending more time working with the data than they ever did before. For example, statistics show that the number of visits and page views is growing faster than the number of unique visitors, implying that users are spending more time on the site.

FIGURE 6-2 Web site visitors and bandwidth.

Then there is the question of what are valuable data and what are not? Figure 6-3 provides an example related to H1N1 pandemic data. Effectively, associated PDB data were hardly used at all but suddenly, during that pandemic, the data became highly accessed. We cannot really tell when data are going to be valuable, so, how do we decide what to keep and what not to keep?

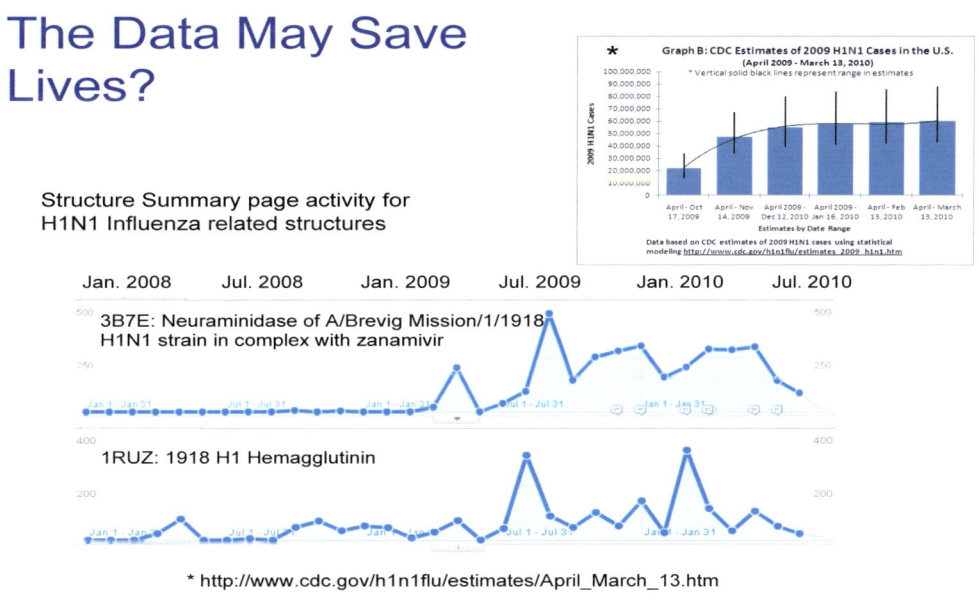

FIGURE 6-3 Data related to the H1N1 pandemic.
Source: Centers for Disease Control.

Let me now talk about data attribution and citation. We spend about 25 percent of the PDB's budget on remediating data that we already have. This has introduced issues related to our support of multiple versions, which we also have handled. There is the so-called copy of record, which is the version that actually went with the publication which we also always make available. The community of depositors of the data also requested that scientists cannot publish their articles unless the supporting data are deposited.

We do cite DOIs but, as I said earlier, no one uses them. Database identifiers are preferred. This exemplifies the kind of problems that we are facing and I think it is a sociological matter. Take the example below of a molecule. It is a receptor on the cell surface called CD4. It is important for many reasons, but one of them is that there is a protein on the HIV virus that binds to it and so it has been identified as the causal agent of HIV infection. When this structure came out, we were not using DOIs.

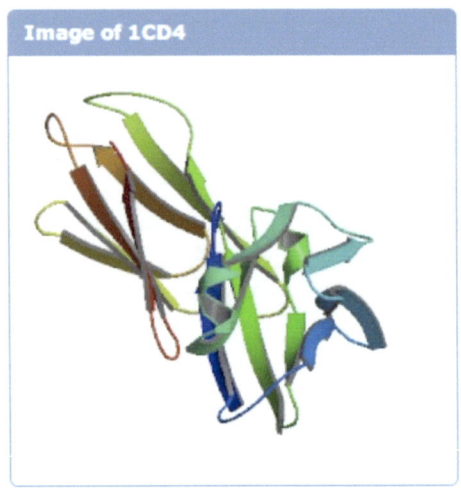

FIGURE 6-4 Image of 1CD4.

Figure 6-4 is known in the biology community everywhere as CD4 because this is the document identifier that it got when it was assigned by a group of scientists at Yale University. That is the identifier we use for character identifiers in the PDB. So what happened is that a group of scientists from Columbia University did a better job on this dataset and then called it in the database 2CD4. The Yale group then went back and did some more work. As a result, they created another dataset and called it 3CD4. But the problem was that 3CD4 had already been given to someone else.

This actually caused angst in the community and it has absolutely no relevance, whatsoever. It is just an identifier.

If there were different sources of those data and we could not be clear on what the copy of record was, it would be a problem but, in this case, it is quite clear. So, rather than thinking about data and journals, it is the idea of trying to bring all this together into a seamless connection. The journal is just one view on the data in some ways and we have been working on this with a number of journals.

Previously, when people did not cite the data, the authors could not use a standard mechanism to find out who has been using the dataset. Now they can because we scan all the open access literature. We can point out to people all of the references that a particular piece of data has in the open access literature. What is even more interesting is that we can then see in each of those articles what else is being cited.

7- Data Citation in the Earth and Physical Sciences

Sarah Callaghan[1]
Rutherford Appleton Laboratory, United Kingdom

When I was asked to speak about the physical and earth sciences, I thought this was a very broad area to cover! So I decided that the best approach was to focus in on a number of issues and examples.

I am a member of the British Atmospheric Data Center (BADC) and we are one of the United Kingdom's National Environmental Research Centre's (NERC's) data centers. NERC funds the majority of the earth sciences and ecological research work in the United Kingdom. I am part of a federation of data centers, which covers the environmental sciences broadly, including hydrology, atmosphere, ecology, ocean and marine, and so on. We deal with a lot of data from many different fields.

It is important in our work to define what a dataset is for ourselves because otherwise, datasets can get very fuzzy. We define a dataset as a collection of files that share some administrative and/or project heritage. In the BADC we have about 150 real datasets and thousands of virtual datasets. We have also 200 million files containing thousands of measured or simulated parameters. The BADC tries to deploy information systems that describe those data, parameters, projects and files, along with services that allow one to manipulate them. Also, in 2010 we had 2800 active users (of 12000 registered), who downloaded 64 TB of data in 16 million files from 165 datasets. To put that into context, less than half of the BADC data users or consumers are atmospheric science users. We have people coming to us to download data for all sorts of reasons, even including school children.

So, what are data for us? Data can be anything from:

- A measurement taken at a single place and time (e.g., water sample, crystal structure, particle collision)
- Measurements taken at a point over a period of time (e.g., rain gauge measurements, temperature)
- Measurements taken across an area at multiple times by a static instrument (e.g., meteorological radar, satellite radiometer measurements)
- Measurements taken over and area and a time by a moving instrument (e.g., ocean traces, air quality measurements taken during an airplane flight, biodiversity measurements)
- Results from computer models (e.g., climate models, ocean circulation models)
- Video and images (e.g., cloud camera images, photos and video from flood events, wildlife camera traps)
- Physical samples (e.g., rock cores, tree ring samples, ice cores)

---

[1] Presentation slides are available at http://sites.nationalacademies.org/PGA/brdi/PGA_064019.

Historically speaking, even though it was very labor-intensive to create new datasets, it was often (relatively) easy to publish the data in a visual form, like an image, graph or table. This picture is an example of one of the earliest published datasets. It was created by Robert Hooke and dates back to 1665.

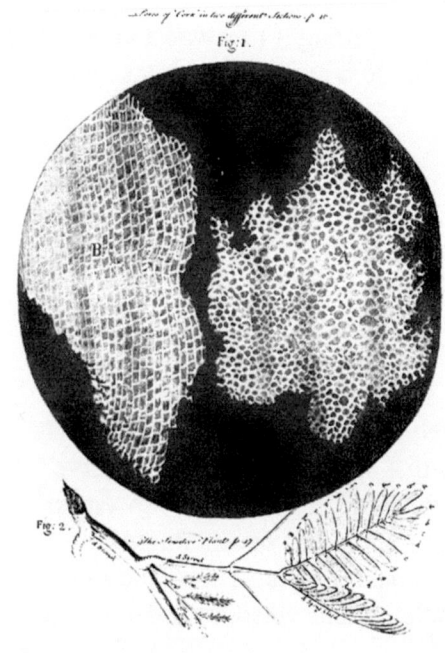

FIGURE 7-1 Suber cells and mimosa leaves.
SOURCE: Robert Hooke, Micrographia, 1665.

One of the big drivers for data citations in the earth and physical sciences is to make it easier to identify products and projects when one is comparing them.

A major example of this is a set of experiments being done by climate modelers all over the world under the auspices of the World Meteorological Organization (WMO) via the World Climate Research Program (WCRP). It is called *CMIP5: Fifth Coupled Model Intercomparison Project.* These climate model experimental runs will produce the climate model data that will form the basis of the fifth assessment report for the Intergovernmental Panel on Climate Change (IPCC). In particular, CMIP5 aims to:

- Address outstanding scientific questions that arose as part of the AR4 (the most recent IPCC assessment report) process,
- Improve understanding of climate, and
- Provide estimates of future climate change that will be useful to those considering its possible consequences.

The method used in CMIP5 is based on a standard set of model simulations which will:

- Evaluate how realistic the models are in simulating the recent past,

- Provide projections of future climate change on two time scales, near term (out to about 2035) and long term (out to 2100 and beyond), and
- Understand some of the factors responsible for differences in model projections, including quantifying some key feedbacks such as those involving clouds and the carbon cycle.

Climate models are usually run on supercomputers, and produce a lot of data. For example, the numbers for CMIP5 are below:

Simulations:
~90,000 years
~60 experiments
~20 modeling centers (from around the world) using
~30 major model configurations
~2 million output "atomic" datasets
~10's of petabytes of output

Of the replicants:
~ 220 TB decadal
~ 540 TB long term
~ 220 TB atmosphere-only
~80 TB of 3hourly data
~215 TB of ocean 3d monthly data
~250 TB for the cloud feedbacks
~10 TB of land-biochemistry (from the long term experiments alone)

These numbers are not particularly important from the point of view of data citation, but they do indicate the sheer volume of data that has to be dealt with. It is not only climate scientists who will have to work with these data, but members of the general public will also try and make sense of them. This is the sort of data that will impact how governments will plan for the next 10 to 50 years.

The researchers who are supporting the whole CMIP5 data management effort have spent a great deal of time and effort thinking about and preparing for how they can store and manage the data. Quality control of the data is also important, not only to ensure that valid cross-comparisons between model runs can be made, but also because this is important to the data provenance and it provides reassurance to the outside world that the data are not being deliberately hidden or obfuscated. CMIP5 (and the climate modelling groups involved in it) will continue to produce a lot of data! It is an international effort, with everyone involved wanting to ensure proper citation, attribution and location of the data produced. Citation will allow the researchers to have traceability and accountability for their datasets.

CMIP5 has issued the following guidelines for the citation of datasets (quote is from the CMIP5 website):

> Digital Object Identifiers will be assigned to various subsets of the CMIP5 multi-model dataset and, when available and as appropriate, users should cite these references in their publications. These DOI's will provide a traceable record of the analyzed model data, as tangible evidence of their scientific value. Instructions will be forthcoming on how to cite the data using DOI's.

At the BADC, we have for many years now had a citation approach where in all our dataset catalogue pages you will find a little box which gives the proper way to cite that particular data set. We have attempted to produce some metrics on how many people actually used these citation instructions, unfortunately without great results. I think this is because users of our datasets do not have the culture of citing data in the first place.

That is something we need to change. We are currently working with all the other NERC data centers to assign DOIs to certain datasets that meet our technical criteria. We expect that this will make it more obvious to our users what the correct way to cite a dataset and use a DOI is, and will encourage more of our users to use the citations.

In terms of earth sciences, the Pangaea data center (http://www.pangaea.de) is further ahead than us when it comes to assigning DOIs to data sets. If you look at their repository catalogue pages they give the citation for the dataset with the DOI and then it says, "supplement to", which gives the citation for the paper of reference.

Finally, I work at the same site as ISIS, which is pulsed neutron and muon source produces beams of neutrons and muons that allow scientists to study materials at the atomic level using a suite of instruments, often described as 'super-microscopes'. It supports a national and international community of more than 2000 scientists who use neutrons and muons for research in physics, chemistry, materials science, geology, engineering, and biology. ISIS is now issuing DOIs for experiment data to allow easy citation. Principal investigators will be sent DOIs shortly before their experiment is due to start. DOIs issued by ISIS are in the form of: 10.5286/ISIS.E.1234567. The recommended format for citation is: Author, A N. et al; (2010): RB123456, STFC ISIS Facility, doi:10.5286/ISIS.E.1234567

Let me conclude by saying that the flood of data is now so great that scientific journals cannot now communicate everything we need to know about a scientific event, whether that is an observation, simulation, development of a theory, or any combination of these. There is simply too much information, and it is too difficult to publish it in the standard journal paper format. Data always have been the foundation of scientific progress—without them, we cannot test any of our assertions. We need to provide a way of opening data up to scientific scrutiny, while at the same time providing researchers with full credit for their efforts in creating the data.

We need data citation not only to provide credit to the scientists who create data, but also for the general public to provide traceability and accountability and to show that as far as possible, we are doing our jobs the way we should. Also, there is serious pressure in the earth and climate sciences to publish data, but there is also a need to ensure proper accreditation. Finally, how we communicate scientific findings is changing and data citation practices are a big part of that.

8- Data Citation for the Social Sciences

Mary Vardigan[1]
University of Michigan, Inter-university Consortium for Political and Social Research

This presentation focuses on norms and scientific issues in the social sciences, and their implications for data citation and attribution. Social science advances like other data-driven disciplines through knowledge claims presented in the literature. Secondary analysis, which enables other scientists to extend and to verify those claims using the data, is an important component of social science research. It is expensive to collect data and it is usually the case that the analytic potential of a dataset is not exhausted by the original researchers, so making data available for others to use makes good sense. To be used by others, data need to be shared and discoverable with proper attribution, and thus there is a need for good data citation practice.

Data sharing

A strong tradition of data sharing, both formal and informal, exists in the social sciences. There are many active social science data archives around the world that disseminate research data; an example of a successful data archive in this space is the Inter-university Consortium for Political and Social Research (ICPSR), which is one of the largest such archives and has been in existence since 1962 with the goal of making data available to others for reuse. Some social scientists request funding to distribute their data through Web sites designed for data dissemination. Despite all of the data sharing activity in the social sciences, Pienta, Alter, and Lyle (2010)[2] have found that about 88 percent of data generated since 1985 have not been publicly archived.

Metadata

Metadata are critical to effective social science. A typical social science data file in ASCII format appears as a matrix of numbers, requiring technical documentation—often referred to as a codebook or metadata— to understand what the numbers represent so that the data may be interpreted. Metadata may also be found in other forms such as questionnaires, user guides, methodology descriptions, record layouts, and so forth.

In general, all of this metadata and documentation in the social sciences is quite heterogeneous in format and most of it is unstructured. The Data Documentation Initiative (DDI) is an effort to create a structured machine-actionable metadata standard for the social sciences. This effort is gaining traction and is used increasingly by data archives and major data projects around the world. It is important to acknowledge the critical role of metadata because when we are citing data, we are implicitly also citing the documentation that is used to understand the data.

---

[1] Presentation slides are available at http://sites.nationalacademies.org/PGA/brdi/PGA_064019.
[2] Pienta, Amy M., George C. Alter, and Jared A. Lyle (2010). "The Enduring Value of Social Science Research: The Use and Reuse of Primary Research Data." http://deepblue.lib.umich.edu/handle/2027.42/78307.

Granularity and versioning

Granularity and versioning of datasets are both important in the social sciences. Social science studies may be single datasets or aggregations. For instance, a longitudinal study may include several discrete datasets, one for each wave of data collection. ICPSR provides data citations at the study level but other data providers are citing at the dataset level.

There is also a need in the social sciences to cite deeper into the dataset. Articles often include and it is important to understand exactly which data are behind those tables.

Data in the social sciences are sometimes updated, so there is a need for versioning to indicate corrections or the addition of new data.

Types of data

In general, ICPSR and its sister archives around the world hold mostly quantitative data, both micro-data and macro-data, but qualitative data are increasingly being generated and archived. We are also seeing that the boundaries between social sciences and other disciplines are blurring. Social science and environmental data are being used together to yield new findings. Survey data are being supplemented by biomarkers and other biomedical information and are being merged with administrative records to provide richer information about respondents. In general, there is a trend towards greater complexity because funders are supporting innovative collections that are multi-faceted, rich, and comprehensive. Social media data and video and audio data are also being used.

Disclosure risk in data

Preserving privacy and confidentiality in research data is a key norm in the social sciences. Survey respondents are promised at the time of data collection that their identities will not be disclosed, and the future of science depends on this ethic.

Providing access to archived confidential data must be done in the context of legal agreements between the user and the distributor. New mechanisms for analyzing restricted data online are coming into existence—for example, we are seeing virtual enclaves and synthetic datasets. There are online analysis systems that enable the user to analyze restricted-use data with appropriate disclosure risk protections, such as suppressing small cell sizes.

It is often the case that a public-use version of a dataset may coexist with a restricted-use version that has more information on it—more variables, and possibly more information about geography. These versions need to be distinguished. This has implications for data citation.

Replication

Replication is, of course, important for science in general. Most claims in the social science literature cannot be replicated given the amount of information that is provided in publications. The community has been working to remedy this situation. ICPSR has a publication-related archive, a small subset of its holdings that is intended to be a repository for all the data, scripts, code, and other materials needed to reproduce findings in a particular publication.

It is important to understand the chain of evidence behind the findings and to have some idea of the record of decisions made along the way to the final analysis. Sometimes this is called deep data citation and provenance. We need both production transparency (i.e., how the data are transformed to get to the final analytic file) and then transparency about how conclusions were drawn.

Data citation practice

There is some tradition of data citation in the social sciences. A standard for citing machine-readable data files was created by Sue Dodd in 1979, and ICPSR has been using a variant of that standard. The Census Bureau has also been providing citations since the late 1980s. Journals are beginning to cite data in a way that is useful. We have found through Google Scholar that some of ICPSR's citations have actually been used. In the social sciences, persistent identifiers for data are now being assigned. ICPSR uses DOIs, but handles and Uniform Resource Names (URNs) are also used.

With respect to journal practices, historically there has not been much effort put into citing data properly or in the right place in articles. There is, however, a growing movement and a lot of momentum behind good data citation practice now. Many publishers are requiring that the data associated with publications be publicly available. In fact, the *American Economics Review* states that they will publish papers only if "data used in the analysis are clearly and precisely documented and are readily available to any researcher for purposes of replication."

At ICPSR, we have been working with our partners in the Data-PASS project to influence journal practices in this area. Data-PASS, or Data Preservation Alliance for the Social Sciences, is an alliance of social science data archives in the United States, including the Odum Institute, the Roper Center, Harvard's Institute for Quantitative Social Science, the University of California at Los Angeles, and the National Archives and Records Administration. Data-PASS has mounted a campaign to contact the professional associations that sponsor journals. We have written to them highlighting the inconsistencies in their data citation practices and have had some success. The *American Sociological Review*, for example, has changed its submission guidelines to require data citations in the reference section of articles and to require persistent identifiers for data.

Linking data and publications

Linking data and publications is also important in the social sciences, just as in the natural sciences. When data citation works as it should, these linkages will happen in an automated way, but up until now, linking data and publications has been a manual process. ICPSR has developed a bibliography of over 60,000 citations to publications that use ICPSR data, with two-way linking between the data and the publications. Vendors like Thomson Reuters are also interested in these linkages.

Summary

In summary, some of the key issues for the social sciences include versioning, which is important for the social sciences because archived data can change substantively over time with new additions. Granularity is an interesting issue as well; it would be useful to define and publish best practices and guidelines for the granularity of data objects that we intend to cite. There is a need to identify very small units like variables uniquely in social science data; they do not need full citation, but identifying them in a globally unique way is important. Metadata seems particularly significant in the social sciences because there needs to be a durable link between the metadata and the data. And, finally, there is replication, a key tenet across the sciences. It is critical that we cite and provide access to all the information necessary to reproduce findings.

It is encouraging that many are thinking about these issues across domains and are working on technological solutions to several of the problems identified.

9- Data Citation in the Humanities: What's the Problem?

Michael Sperberg-McQueen[1]
Black Mesa Technologies

As a complement to the presentations on data citation in the life, natural, and social sciences, this presentation will discuss data citation in the humanities. The reaction of some people to that topic will be to say: "What?! When did humanists start working with data? What is considered to be data in the humanities?" To respond to these questions, I will start with a little background on what counts as data in the digital humanities, then survey current practice with respect to data citation in this domain, concluding with some remarks on intellectual issues with data citation in the humanities.

Data in the digital humanities

Humanities scholars started using machine-readable data in 1948, when Father Roberto Busa began work on the *Index Thomisticus*. This index was a full-text concordance of every word of every work published by St. Thomas Aquinas, as well as every work that historically had ever been attributed to Aquinas (even those that Busa felt confident were not actually by Aquinas). In his theological research, Busa was particularly interested in the concept of the eucharistic presence, which required an exhaustive list of Aquinas' use of the preposition *in*. Since prepositions tend to have very high frequency in natural-language texts, such a list is very hard to prepare with note cards.

During the 1950s and 1960s, a great many individual texts were encoded and work of various kinds was performed with them. Often, concordances were made; sometimes stylistic studies were performed. In the early 1960s, Henry Kucera and Nelson Francis created the Brown Corpus of American English, a one-million-word corpus drawn from works published in 1961.

The journal *Computer and the Humanities* began publication in 1966, and in the 1970s, organizational structures formed to support the field: the Association for Literary and Linguistic Computing (1973) and the Association for Computers and the Humanities (1978). Also in 1978, the Lancaster-Oslo-Bergen Corpus of British English was published, created as a *pendant* to the Brown Corpus of American English.

Typically, because of the amount of work involved, the first texts to be encoded were the sacred texts. Busa, as a theologian, found Aquinas worth the effort of 30 years of work (the *Index Thomisticus* was finally published in 1978), and the Bible too was one of the earliest texts put into electronic form.

Of course, the classics were not far behind; classicists are accustomed to very labor-intensive work with their texts. The *Thesaurus Linguae Graecae* project was started in the 1970s with the

---

[1] Presentation slides are available at http://sites.nationalacademies.org/PGA/brdi/PGA_064019.

aim of making a comprehensive digital library of Greek writing in all genres from Homer to the fall of Constantinople in 1453; it issued its first CD-ROM in 1985.

What counts as data in the humanities? The humanistic disciplines seek, in general, better understanding of human culture, which means that pretty much anything created by humans is a plausible object of study. Data in these fields may include:

- digitized editions of major works;
- transcriptions of manuscripts;
- thematic collections (e.g., author, period, genre);
- language corpora (balanced or opportunistic; monolingual or multilingual [parallel structure or parallel-text translation equivalents]);
- images of artworks (e.g., Rossetti, Blake, DeYoung Museum ImageBase); and
- maps.

Digital representations of pre-existing artifacts now often take multi-media forms, e.g., scans plus transcriptions. Human culture did not end when humans built computers, however, so digital artifacts and born-digital objects are also objects of study for humanities disciplines. Scholars are studying digital art forms, hypertexts, interactive games, databases, and digital records of any kind. There is no human artifact that is *a priori* unsuitable as an object of historical or cultural critical study.

If humanists have been creating large, labor-intensive, expensive digital resources for six decades, the question arises: is anyone taking care of the materials thus created? The answer is yes and no.

Publishers of digital editions presumably have a commercial interest in retaining an electronic copy of the edition. (I do not actually have any evidence that they are aware of having that interest or that they are acting on that interest, but I hope some of them are, because to the extent that publishers regard digital editions as commercial products, they have historically not wanted to deposit the editions with a repository for long-term holding.) Individual projects also have an interest in the preservation of the materials they create, but individual projects sometimes suffer from the illusion that they are going to live forever, with the consequence that they are often taken by surprise when their funding runs out. So neither publishers nor individual data-creating projects have, as a rule, been reliable long-term custodians of humanities data.

From as early as the 1970s, there have been projects to collect electronic texts and to preserve them in an archive, beginning with Project Libri at Dartmouth University, which no longer exists. Fortunately, before Project Libri was terminated, its managers deposited all their texts with the Oxford Text Archive, founded a few years later in 1976, which does still exist. There are currently many electronic text centers in university libraries, which have retained digital objects. As a rule, however, most library-based electronic text centers are interested in displaying and making accessible things that they have digitized and are less interested in acquisition of digital resources from elsewhere. Also, there are (at least in theory) digital repositories of various kinds, including institutional digital repositories.

Although there has been a long history of this work, there is nothing in the humanities like the network of social science data archives that we see with the Inter-university Consortium for Political and Social Research (ICPSR) in the United States, the Economic and Social Research Council Data Archive in the United Kingdom (UK), or the Danish Data Archive, and their various analogues in other countries. The Arts and Humanities data service in the UK essentially tried to fill that role and did so very successfully for the eleven years that they were funded before they were terminated. (It should be noted that the ESRC Data Archive has now been renamed the UK Data Archive and describes itself as holding research data in "the social sciences and humanities."

Current data citation practice in the humanities

The current status of data citation practice in the humanities is mixed. There are some hopeful signs. For example, the TEI Guidelines explicitly require internal metadata for the electronic object itself, and not just for the exemplar of the electronic object. So in principle digital humanists should be familiar with the idea that an electronic object representing a non-digital artifact is distinct from its source and needs to be documented and cited in its own right. In theory, at least, people should know what to cite.

Also, most citation styles used by humanists now have been revised to allow the inclusion of IRIs (internationalized resource identifiers) in the citation, which ought to be helpful. Some citation styles, of course, refer not to IRIs but to URIs (uniform resource identifiers) or URLs (uniform resource locators) instead, which will irritate those of us who believe IRIs should be the identifiers of choice. But the principle of providing an identifier for an electronic object is at least recognized.

In preparation for this presentation, I examined a random sample of papers in the field, looking to see whether there are fairly complete digital resources behind these papers, and if so, whether I could understand or read what they are? There are, as might be expected, several patterns. Ideally, one might want to see published resources used in the papers, with explicit citations in the references so services like ISI will find them and so that the data citations will show up in a citation index. As far as I can tell, however, this is purely a theoretical category: I found no instances of it in my small sample.

The closest thing found in the sample to this ideal practice were papers which mention published resources, which are explicitly described, sometimes with a URL pointing to the item, but with no reference to the resource in the references. Sometimes, the references include instead a reference to a related paper, which may indicate both a desire to cite the work and a discomfort with citing resources which do not take traditional scholarly forms, or perhaps uncertainty about how to cite data resources directly.

In other cases, papers mentioned resources that are clearly identifiable as objects, that clearly have an identity of their own, and that have not been published. The resources are explicitly mentioned and acknowledged in the text, but naturally enough they are not cited because they have not been published. There is not even the equivalent of a personal-communication citation.

There also are many resources that clearly must exist unless the paper is a complete forgery, but they are unpublished. These resources are implicit in the argument of the paper, but they remain cited because, clearly, the author thinks of them as analogous to working notes.

In summary, the situation for data citation in the humanities is completely confused.

Some problems in humanities data citation

Some of the problems arising for data citation in the humanities are problems already discussed in the natural and social science context. Others may be particular to the humanities.

If I, as an author, want to cite a resource, how should it be cited? What exactly should be cited? Am I citing the entire British National Corpus? Am I citing a particular sample of the British National Corpus? Am I citing the archive from which I got the British National Corpus? In the case of the British National Corpus, those distinctions are very clear. In other cases, the distinction is not clear at all. When working with digital resources which combine and recombine with each other in unpredictable ways, scholars will find philosophical questions of identity taking on an unwontedly urgent practical aspect.

Second, it is sometimes difficult to locate reliable metadata concerning a resource one might want to cite. Without a physical title page, it may be challenging to identify the title of a data resource, or the names of those intellectually responsible for its content, or the nature of their contribution. In many cases, it is difficult to identify a publisher or a date of publication. Those responsible for distributing a data resource may not regard themselves as the publishers of the work, because they do not regard themselves as engaged in publishing in the conventional sense. (This problem may be familiar to those who have tried to cite microfilms created for individual scholars by individual photographers.) It is not even clear whether familiar roles like "publisher" have the same relevance for electronic resources as they do for print materials. If the familiar division of labor among publishers, distributors, repositories, libraries and archives is, fundamentally, a way of organizing the management of information, we may expect those roles to be recognizable in the digital world. If, on the contrary, that familiar division of labor reflects only a way of organizing the management of paper and other physical objects, then the digital world may well converge on a different and incommensurable set of roles.

In many ways, the challenge of locating reliable metadata among them, digital objects seem to be in an incunabular phase; like the earliest printed books, digital objects lack established conventions for identifying the object or those responsible for it. In pessimistic moments, an observer might fear that the situation is even worse than that, and that the creation and dissemination of digital objects does not resemble the creation and dissemination of early printed books so much as it resembles scribal transmission of manuscripts.

A third challenge for citation of humanities data resources is that many of those involved, whether as producers or as consumers of resources, want turn-key systems, not a set of tools and materials which leave them to their own devices. This perfectly understandable desire for ease of use tends in practice to lead to tight coupling (both in the technical sense, and psychologically in producers and consumers) of the data resource, the software used to provide access to the resource, and the user interface of that software. (There are notable exceptions, including the

Perseus Digital Library, which has over the years released its data with different software, as the computational infrastructure available to its users has changed.) When a data resource is tightly coupled with a particular piece of software, it can be difficult to distinguish the one from the other, with consequent difficulties for anyone who would like to cite the data resource itself, and not the data-plus-software combination.

Several factors may be identified which tend to inhibit data citation in the humanities.

Fear of copyright issues: When digital resources are constructed without copyright clearance (perhaps on the theory that they are for personal use only), the creators of the resources will understandably hesitate to publish them, or even to cite them explicitly. As a senior figure in the field wrote to me: I think you will still find plenty of people saying "we ran a stylometric analysis on a corpus which has these properties, but we cannot let you see the actual corpus because we did not obtain the copyright."

Anti-scientism: Citing data resources may seem foreign to the culture of humanistic scholarship, an eruption into the humanities of natural-scientific practices and perhaps a symptom of science envy, to be discouraged as naïve and unhelpful.

Citation chains: Print has (reasonably) well-established conventions for re-publication and citation of earlier publications. Not so digital resources, which may include refinements, revisions, elaborations, subsets, derivations, annotations, and so on, often made without any explicit reference to the sources from which they were derived. This is not unknown in print culture, of course: some editions of classic authors provide no information about the copy text used in the edition. But it does make it hard to trace the provenance of some digital resources, and when resource creators fail to cite the prior resources they use, it is not surprising if users of the later resources also fail to cite the resource when they use it.

Versioning: Large humanities projects typically make multiple passes over the same material. In the future, it is not unlikely that early results will be published (under pressure from the Web culture and from funders). If there are multiple versions of a resource from the same source at least two problems may be expected.

Will the metadata for the resource label the version and explain the nature of its relation to other versions of the resource?

What will be the unit of change? Will changes be clumped into groups in the way familiar from print editions? Or will the resource change continuously (in which case, will it be possible to pluck out a given state of the resource at a given instant to represent a version of the resource?

Quiddity: Large humanities projects typically make multiple passes over material.
reading text; text-critical variorum text; text with literary annotations; linguistic annotations (glosses for cruxes? parse trees? ...); or formalization of propositional content.
Which of these is the thing I am publishing? And which of these is the thing I am citing?

Longevity: Finally, there is the question of longevity. It is well known that the half-life of citations is much higher in humanities than in the natural sciences. We have been cultivating a culture of citation of referencing for about 2,000 years in the West since the Alexandrian era.

Our current citation practice may be 400 years old. The http scheme, by comparison, is about 19 years old. It is a long reach to assume, as some do, that http URLs are an adequate mechanism for all citations of digital (and non-digital!) objects. It is not unreasonable for scholars to be skeptical of the use of URLs to cite data of any long-term significance, even if they are interested in citing the data resources they use.

DISCUSSION BY WORKSHOP PARTICIPANTS

Moderated by Herbert van de Sompel

PARTICIPANT: This is a question for Mary Vardigan. When you have data systems that are based on surveys, do you have to include in the metadata how exactly the survey was conducted? The reason I am asking is that a couple of years ago, my city paid a contractor to do a big survey of citizen satisfaction. It was done classically with randomly drawn phone numbers and when I looked into it, there was no list of people with cell phones. As a result, cell phones, essentially used by the Hispanic population, were vastly under-sampled. How do you deal with this kind issue?

DR. VARDIGAN: That is a good question. The sampling information is part of the important documentation and metadata that we distribute with every dataset we make available. It is important in assessing data quality. At ICPSR, we do not assess data quality ourselves; it is the community that will determine whether the sample is adequate and scientifically sound. It is important, therefore, to have that descriptive information about how the survey was conducted.

DR. BORGMAN: We were taught in this session about what might be generic solutions across the disciplines, as well as what could be specific. So far, I have heard two things in common across them. One is that there are data papers or surrogates, where a journal article that describes the data will be cited in lieu of citing the data per se. The other is that there is a deep complexity and confusion in the field. I think it would be good if each of you could highlight what you heard from the other panelists that might work in your field and might be really useful.

DR. SPERBERG-MC QUEEN: Another item to add to your list is the issue of granularity and the perceived need to be able to cite parts as well as entire datasets.

DR. CALLAGHAN: I will emphasize the importance of metadata when it comes to doing data citation. It is not enough to validate our datasets unless we have got the full description of what the numbers are and what they actually mean.

DR. VARDIGAN: I would like to add the issue of versioning, which seems to be common across disciplines, because data do change over time. There are some dynamic datasets that are continually being generated and there may be discipline-specific solutions to this issue that could be deployed across other disciplines if we knew more about them.

DR. BOURNE: The notion of peer review of data is being brought up in different contexts. My sense is that this is something that is dependent on the maturity of the field that is generating those data. Unlike new fields, when a field is pretty mature, I think the community must have come to a good understanding of how the data could be peer reviewed as part of the process.

DR. CHAVAN: I want to emphasize that some of the data that we deal with are very complex and this affects our ability to cite them properly. What we have done is that we got involved with a publisher who is innovative and willing to experiment to work on this issue. GBIF currently publishes more than 18,000 datasets and none of those datasets have more than four metadata fields complete out of 64 lines. So, publishing data is easy, but writing metadata is really a difficult task.

We came up with the approach of publishing the metadata document as a scholarly publication. We publicized an announcement about three months ago regarding a technical solution and a recognition mechanism that we will put in our papers. In the last three months, we approached about 350 data publishers who publish 18,000 datasets and few of them took it up. They said that writing a good metadata document that can actually be published is difficult because every metadata document will have to go to a review process. This is difficult and I personally did it. I wrote a metadata document that could qualify for review and it took about eight hours. This is something that needs to be addressed. Nonetheless, out of the 350 publishers, we were able in three months to convince about ten of them.

DR. CALLAGHAN: I agree that comprehensive dataset management and metadata work are very challenging tasks. However, we are in the position of having good incentives for data producers until they give us their metadata. Metadata are very important to help data producers understand the complexity of the systems related to data management. When it comes to the peer review of metadata, I do not think that we want to get scientists and journals involved in this process because it is time consuming, complicated, and technically biased toward data management professionals. I am of the opinion that it is the job of the data centers to make sure that the metadata are complete. We can do that and it is well within our area of expertise.

DR. VARDIGAN: Just one more point related to incentivizing data producers to create good metadata. In the United States, we now have the National Science Foundation asking researchers applying for grants to provide data management plans, and metadata are a big component of those plans. We are hoping that this will be a positive influence on what eventually gets deposited into the data centers.

DR. BOURNE: I think that the provision of good metadata is dependent on the reward systems. In some biosciences communities, it is not only that you cannot publish without depositing the data, you must also deposit that data with a fair degree of rigorous metadata. That is also a reward because you cannot publish without it.

PARTICIPANT: I want to ask all the panelists how do you see the chances of some entity that would be a registry of unique and persistent identifiers across the different domains?

DR. BOURNE: Can you turn it around? The problem right now is that publishing is a competitive business and no single publisher is going to demand getting something standardized, because there is a risk to their business model. However, if publishers got together and insisted that there should be a standardized metadata description that we can use across the board, there could be a chance for it to happen.

DR. CALLAGHAN: I would say that DOIs are very well accepted and that is the route that we have chosen to use as far as our datasets are concerned. I would like to add, though, that a DOI should not be the sole basis of the citation. There has to be more information on the DOI because a DOI is just an alpha-numeric string. A person will look at it and will not understand anything.

Whereas, if you have another part of the citation that gives the author name, title, and perhaps other information it might not be any good for computers, but it will help the humans.

DR. BRASE: I think part of the issue about how much metadata is already contained in an archive depends on the discipline. On the one hand, for example, in the life sciences, there is a large amount of data already in their archives and they have their own way of doing things. Therein lies the problem. Change will be difficult. On the other hand, for long-tail data that do not have a home, I think there is real opportunity because people want to deposit those data and get credit for them. This is where you can standardize the process. That is where DataCite and other similar initiatives can come in.

DR. WILSON: It is important in a lot of fields that we have documentation of the method by which the data was generated. I also want invite more comment on the division between what are metadata and what are data, because this is not always a clear line—one person's data may be another's metadata, and vice versa.

DR. VAN DE SOMPEL: I agree. I think that we still have a huge gap in our understanding of what people consider to be "data."

PARTICIPANT: I would like to see a universal approach or guidelines in relation to data citation and attribution. I think that it is very encouraging that we have people representing different fields and areas looking at the same set of problems here. Let us just try to be simple and work on things gradually. These discussions and emerging tools and technologies have tremendous motivation for publishers and researchers, and I think that this meeting is a very good starting point. We might not come up with the best solution right now, but as time goes on, I think it is very encouraging.

PARTICIPANT: Learning from each other is a useful approach as well. I have been working towards citable references for a long time and as I had this subject as a priority; it has been less of a priority in other disciplines. Other disciplines suffer from this syndrome of incrementally refined datasets. For example, sequences from GenBank are refined by many people and that makes it a complicated co-authorship situation. Have you run into that in ICPSR and, if so, how do you handle such a situation in a cited reference?

DR. VARDIGAN: We have some instances of data that have multiple contributors. Some datasets have over a hundred contributors. We have just used "Name, et al." to acknowledge the variety of people involved. We do not have a specific approach to deal with such a situation.

PARTICIPANT: I would be very interested if Dr. Vardigan or other colleagues can talk about third-party metadata. For example, if there is a record put somewhere with an appropriate link that states, "this sample did not include cell phones," this would tell us that the sample is biased. That would be a really useful approach to have.

DR. VARDIGAN: I do not know of anything like that currently in existence, but we all rely on the scientific method. If a paper is published and others decide to make judgments about its merits and publish something themselves about the quality of the data or its content, they can do that. As a data center, ICPSR does write what we consider to be comprehensive metadata references, and we track publications based on our data.

PART THREE

LEGAL, INSTITUTIONAL, AND SOCIO-CULTURAL ASPECTS

10- Three Legal Mechanisms for Sharing Data

Sarah Hinchliff Pearson[1]
Creative Commons

Sharing data today can be easy; you can simply post them on the web. But doing so means losing some control over the data, including whether you will be accurately and properly credited. This is obviously the case when you share data without a related license, contract, or waiver. As I will explain, to a certain extent this is true even when any one of those legal mechanisms is used.

I will begin by defining some terms. For purposes of this presentation, attribution, credit, and citation all have distinct meanings. *Attribution* refers to the legally imposed requirement to attribute the rights holder when the data are copied or reused in a specified manner. The remedy against someone who fails to attribute is a lawsuit, either based on breach of contract or infringement of an intellectual property right, depending on the legal mechanism used to impose the attribution requirements. *Credit,* on the other hand, is what we all want—explicit recognition for our contribution to someone else's work. Finally, there is *citation*, which is rooted in norms of scholarly communication. The purpose of citation is to support an argument with evidence. However, citation has also become a proxy for credit, albeit an imperfect one.

This is an important starting point. It reminds us that legal attribution requirements do not necessarily match our expectations for receiving credit, nor do they perfectly map to accepted standards of citation. When the remedy for failure to attribute is a lawsuit, we are well-served to recognize this incongruity. With that in mind, let us turn to the law.

There are three main legal mechanisms for sharing data: licenses, contracts, and waivers. Whenever data are shared, there is a possibility they will not be properly cited upon reuse. Licenses and contracts attempt to eliminate this risk by imposing legal attribution requirements. Waivers, however, do not legally impose attribution. Instead, they rely on community norms to ensure proper citation. There are consequences to each of the three approaches. I will address each below.

Licenses

We will start with the approach for which Creative Commons is best known - licenses. Licenses operate by granting permission to copy, distribute, and adapt data upon certain conditions. One of those conditions is attribution, as it is in all Creative Commons licenses. A license sounds a lot like a contract because it grants permission to use data under certain conditions. However, they are actually quite different because a license is built upon an underlying exclusive right. Therefore, in order to understand the scope of a license, you have to understand the scope of the underlying right. In the context of sharing scientific data, the rights involved are typically copyright or database rights.

---

[1] Presentation slides are available at http://sites.nationalacademies.org/PGA/brdi/PGA_064019.

We will begin by taking a closer look at copyright law. Copyright law grants a bundle of exclusive rights to creators of original works at the moment the work is fixed in a tangible medium. In non-legalese, that means copyright is granted automatically once you write your work down or enter it into the computer.

Copyright is limited in scope and duration, and the specific limitations vary by country. For scientific data, the most important limitation of copyright is that copyright never extends to facts. Copyright does, however, extend to a collection of facts if they are selected, arranged, and coordinated in an original way. The required threshold is low.

There is significant uncertainty about where the line of copyright extends, even among copyright lawyers. To complicate matters further, this line varies somewhat according to the laws of each country.

Determining what is subject to copyright is only the first hurdle. The next task is identifying the scope of copyright protection. Even when a database or a collection of facts is subject to copyright, the facts themselves remain in the public domain. This means that the general rule in the U.S. and elsewhere is that data can be extracted from a copyrighted database *without* infringing copyright law.

That is not true, however, in the European Union (EU). In the EU and a few other countries, governments have implemented what are called *sui generis* ("of their own kind") database rights. These rights allow a database maker to prevent the extraction and reuse of a substantial part of the contents of a database, even if the contents are otherwise in the public domain.

A license can be built atop copyright or database rights or both. By way of example, Creative Commons ("CC") licenses are copyright licenses. If a CC license is applied to a database, it covers both the data and the database, all to the extent each is subject to copyright. Any use of the data or database that implicates copyright, requires attribution. Any use of the data that does not implicate copyright – if for example, the data are in the public domain – does not require attribution, even if it triggers database rights.

Because of the difficulty of deciphering the contours of copyright protection in scientific data and databases, it is very hard for both the data provider and data user to know when the license applies and when it does not. In other words, it is difficult to know when attribution is legally required. This creates a number of risks.

For one, it creates the risk that data providers will be misled about what they are getting when they apply a license to their data. They may believe that if they apply a license to their data, any use of the data will require attribution. As I explained earlier, that is not the case. If the data are in the public domain, or if the use of copyrighted data falls under fair use, the attribution requirement is not triggered.

It also creates the risk that data users (also referred to as the licensee) will misjudge their attribution requirements because of the difficulty in determining when copyright applies. They may under- or over-comply with the license without realizing it. Either situation can be problematic.

In addition to the legal uncertainty, licenses also create the risk of imposing burdensome attribution requirements. In the science context in particular, projects often rely on data gathered from a variety of different sources. Depending on the licenses used, it is possible that would require attributing each individual or institution that contributed any piece of data to the project. This is a problem we call *attribution stacking.*

This raises yet another potential problem with attribution. Attribution obligations written into a license are, by their nature, inflexible. No lawyer can anticipate every situation in which the attribution requirements would be triggered and account for all of the circumstances in which they will be applied. This can create some absurd situations where, for example, a user or aggregator of data may technically be required to attribute 1000 different data providers, all in the idiosyncratic manner that the rights holder has dictated. Conceivably, the user could do all this and still not satisfy people's expectations for receiving credit or accepted standards of citation.

Contracts

The next legal mechanism for requiring attribution is contract law. Contracts can have different names and take a lot of different forms, but they are often called data use agreements or data access policies.

Unlike a license, a contract does not necessarily require an underlying intellectual property right. Technically, it requires a few legal formalities, including an offer and acceptance. In practice, sometimes that manifests in an online agreement, where the user has to click to accept the terms to access to data. Other times the user is presumed to have accepted the terms by continuing to use the site. If you read those terms, they may require attribution.

Like licenses, contracts suffer from a number of potential downsides. For one, they likely impose confusing obligations on users who get data from a variety of sources, all subject to different user agreements. This problem is even more pronounced with contracts because at least public licenses are somewhat standardized. User agreements are not, which means each data source likely has a different user agreement, filled with legalese imposing attribution and other obligations on users. The consequence is that some data sources may not be used simply because users cannot understand the terms.

Another limit to contract law is that it only binds the parties to the agreement. That may sound obvious, but this is not the case with licenses. If someone obtains licensed data and shares them, the person who obtains them it from that second user is still bound by the conditions of the license. If the data were shared by contract alone, the person who obtained the data from the second user would not be bound by the terms of the contract because they were not a party to the original agreement. In this respect, contracts have a more limited reach than licenses.

In a different respect, contracts have a broader reach than licenses. Because they are not tied to an underlying right, contracts can impose obligations on actions that are not restricted by copyright or database rights. The effect could be to restrict or take away important rights granted to the public. For example, in 2011, the Government of Canada launched an open data portal with a related contract controlling access to the data. This agreement initially had a provision that forbid any use of the data that would hurt the reputation of the Canada. This requirement created an uproar and was changed within a day. Nevertheless, this example shows the potential for overreaching. This sort of thing is particularly troublesome in the context of standardized contracts, where the terms are rarely read and almost never negotiated.

Waivers

The last legal mechanism is the waiver. Waivers can take many forms, but the purpose is to dedicate the data to the public domain.

Waivers are not enforceable in every jurisdiction. To deal with this problem, CC has created a tool called CC0 (read CC Zero) that uses a three-pronged approach designed to make it operable worldwide. The first layer is a waiver of copyright and all related rights. If the waiver fails, CC0 has a fall-back license that grants all permissions to the data without any conditions. As a final backup, CC0 contains a non-assertion pledge, where the rights holder promises not to assert rights in the data.

Obviously waiving rights to a dataset means the provider no longer has control over it. Among other things, that means the data provider cannot require attribution (although they can certainly encourage it). Yet, as mentioned above, nearly every approach requires losing some measure of control in the data. Waivers also provide legal certainty in a way that contracts and licenses do not. There is no need to try to decipher the scope of copyright protection or consult a lawyer. Nor is there a need to try to parse the legalese of a variety of different user agreements. Note this certainty does not exist when data are released without any legal mechanism. The silent approach leaves people guessing about whether property rights exist in the dataset and whether they risk liability by using it.

To summarize, each approach has consequences. With licenses, we face legal uncertainty about the scope of the license, and we risk imposing attribution requirements that are inconsistent with relevant community norms and expectations. With contracts, we gain some measure of legal certainty, but we risk imposing even more burdensome attribution obligations as each institution or data provider creates its own contractual terms. Contracts also pose the risk of overreaching and imposing obligations that may restrict important rights of users. Waivers avoid the problems associated with licenses and contracts, but they require giving up control.

It is important to remember that there is no mechanism that can impose legally binding obligations in a way that perfectly maps to our expectations for receiving credit or accepted standards of citations. By trying to use the law for control, we risk imposing unnecessary transaction costs on data sharing. We also potentially push people away from using our data sources. Choosing the right approach requires an understanding of the consequences. The conversation at this workshop is a good start.

11- Institutional Perspective on Credit Systems for Research Data

MacKenzie Smith[1]
Massachusetts Institute of Technology

My presentation is about the institutional perspective on credit systems for research data. Why does credit matter to the institution? Simply put, it is because academic research institutions depend on reliable records of scholarly accomplishments for key decisions about hiring, promotion, and tenure. These mechanisms evolved over decades for books, peer-reviewed publications, and sometimes grey literature (e.g., theses, technical reports and working papers, conference proceedings, and similar kinds of information that are not peer-reviewed). Also, a lot of services emerged to make assessment of the record easier for the administration. This includes impact factors, academic analytics, and other methods.

The traditional assessment model as we have it now is falling apart because it does not allow new emerging modes of scholarship and scientific communication to be included. For example, the current traditional evaluative process does not consider the following:

- Preprint repositories like arXiv or SSRN (the Social Science Research Network).
- Blogs, websites, and other social media.
- Digital libraries like Perseus, Alexandria.
- Software tools, e.g., for processing, analysis, visualization.

There are important reasons why institutions care very deeply about these issues. One of them is institutional representation. There are national and world rankings in universities. One of the things they look at is the accomplishments of faculty and researchers in the institution. These practices that we come up with, like impact factor, play a big role in some of the ranking decisions, which are extremely important to the administration of the university. Then there is academic business intelligence. Many universities now have major industrial liaison programs and technology licensing offices. They are always trying to figure out what the academics are producing that might be commercialized or otherwise exploited, both for the university's benefit and the researcher. Furthermore, it is important for recruitment. The institution needs to be highly ranked in order to recruit excellent students and faculty members. Finally, there are public relations and fundraising considerations that are extremely important for the university. It is easier to raise money from donors if you have a good reputation and when you have some famous researchers. I know this can be very irritating to those of us who are working in research, but this is real life at the university.

In the past few decades, at least, the publishing process did not really involve the institution at all. The researchers did the research, wrote the papers, and published them on an outsourced basis through their societies, or increasingly with commercial publishers. The university did not get involved until the library bought it back. So, the only role that the university had was as a consumer. The researchers were acting almost like independent agents in that model.

---

[1] Presentation slides are available at http://sites.nationalacademies.org/PGA/brdi/PGA_064019.

However, that is changing with data because in order to produce data, you often need institutional infrastructure. Sometimes it is infrastructure related to a disciple, but a lot of times it is institutional. This is where we get into discipline-related variations. In fields like geophysics and genomics, for example, the infrastructure is not usually provided by the institution, but in the social sciences, it is frequently provided. In the neuroscience field, it is often the institution that funds the various imaging machines and pays for all the storage and infrastructure to maintain the resulting data.

We thus have gone from a system where the institution was not involved in the publishing process to one where the researchers cannot really do what they need to do without support from their institution. Furthermore, institutions have other responsibilities when research is concerned. For example, they have some responsibilities when it comes to funding. The institution is the grantee and is legally responsible for enforcement of the terms of the contract. Also there is additional infrastructure that we all rely on now to do our work, such as digital networks and computing, the library, the licensing office, and the like. The university is responsible for making sure that the infrastructure is well-maintained and functioning. Lastly, institutions are responsible for the long-term storage of scholarly records so they are preserved and will be available and accessible to all interested stakeholders.

Now I need to focus on the intellectual property (IP) part. I would say that to the extent that IP exists in data, or that it has commercial potential, oversight for citation or attribution requirements is unclear (see the presentation by Sarah Pearson). Researchers assume that they control the data and have the intellectual property rights and that they can decide what terms to impose on their data. Often, however, researchers do not, in fact, have these rights. Although funders do not assert intellectual property rights, they frequently do have policies about what should happen to those rights when they give a grant. For example, this is a quote from the NSF Administration Guide: "Investigators are expected to share with other researchers, at no more than incremental cost and within a reasonable time, the primary data, samples, physical collections and other supporting materials created or gathered in the course of work under NSF grants. Grantees are expected to encourage and facilitate such sharing."[2]

Also, university copyright policies are evolving. This is another quote from an unnamed university's faculty policy. "In the case of scholarly and academic works produced by academic and research faculty, the University cedes copyright ownership to the author(s), *except where significant University resources (including sponsor-provided resources) were used in creation of the work*" [italics added].

This quote is typical. You can find a similar formulation in just about every institution's faculty policy document. This is what historically has been applied to things such as software platforms developed with university infrastructure. The same thing is being applied to data now. Note that the word "significant" in the statement is not defined.

Patent policy is similar. Here is another quote from an unnamed university: "Any person who may be engaged in University research shall be required to execute a patent agreement with the

---

[2] NSF Award and Administration Guide, January 2011.

University in which the rights and obligations of both parties are defined." In other words, researchers do not get exclusive rights to their patents. They will have to negotiate with the university. This is somewhat vague, however. When data have commercial potential, and they do sometimes, this starts to get really interesting.

The new NSF requirement was not received well by all researchers. Some said: "I think I might be able to patent something from these data that will make me money. So please keep your hands off my research. I am not sharing." I am exaggerating to make a point here, underscoring the fact that as commercial applications of data become better understood, especially in the life sciences and engineering, this could become a really tricky area for everyone involved in academic research.

From an institutional perspective, some of the requirements for data citation include:

- Persistent or discoverable location
    - Works even if the data moves or there are multiple copies
- Verifiable content
    - Authenticity (i.e., "I am looking at what was cited, unchanged")
    - Requires discovery and provenance metadata
- Standardized
    - Data identifiers: DataCite, DOIs
    - People identifiers: ORCID registry
    - Institutional identifiers: OCLC? NISO I2?
- Financial viability
    - Identifiers cost money to assign, maintain
    - Metadata is expensive to produce

Let me elaborate on these requirements. First, a citation has to be persistent or provide a discoverable location. We need the citation and the discovery mechanism to work, no matter where the database is located. We need some way of proving the authenticity of the data. In other words, I am looking at a URI that is referenced in a research paper. How do I know that the dataset I get to by resolving that URI is the dataset that the researcher was using at the time? That requires discovery and enough metadata.

We also need more standardization in key areas. We have to have identifiers for the data, but we also need identifiers for the people and for the institutions involved. For example, I am involved in the ORCID (Open Researcher and Contributor Identification) initiative, which is looking at ways of creating identifiers for researchers that would be interdisciplinary, international, and portable across time.

Lastly, there is the issue of financial liability. We have to keep these efforts affordable so we can talk about identifiers, be it DOIs or DataCite URIs. I know that there has been contention for using identifiers for data in the past, since if we are talking about a million researchers, that is one thing, but if we are talking about billions of datasets and data points, all of which need unique URIs, that could cost a lot of money.

Also, the metadata is currently very expensive to produce. This has to be done in a partnership between researchers and specialists who are paid to do this kind of work, whether it is in data centers or libraries. We have to involve experts whose job is to worry about quality control and metadata production, and that is also very expensive. So, we have to keep in mind these issues and requirements when we think about data citation techniques.

12- Issues of Time, Credit, and Peer Review

Diane Harley[1]
Center for Studies in Higher Education
University of California at Berkeley

I will be speaking today as an anthropologist who has spent a large part of the last decade thinking deeply about and conducting research on issues of scholarly communication, the future of publishing, and academic values and traditions in a digital age. I am a social science scholar. I am not an advocate for particular approaches nor an administrator or librarian. That being said, I will put my comments today in the context of our research findings regarding the drivers of faculty behavior, the importance of peer review in academic life, and the various incentives and barriers to scholars regarding where and when to share and publish the result of research (including data) over the entire scholarly communication life cycle (not just in final archival publications such as journal articles and books).

First, an important note about our research and methods. Our work is based upon the rigorous qualitative interview, observational, and textual data collected during the six-year *Future of Scholarly Communication Project (2005-2011)*,[2] funded by the Andrew W. Mellon Foundation. More detailed information on our sample population and research design can be found in Harley et al. (2010: 13-15) and Harley and Acord (2011: 12-13); I include more specific references throughout this paper. I refer readers to the "thick descriptions" of 12 disciplinary case studies[3] and the more extensive literature reviews in these publications. In brief, our sample spans 45 elite research institutions, more than 12 disciplines, and includes the results of more than 160 formal interviews with scholars, administrators, publishers, and others.[4] My comments today will be almost exclusively focused on an elite class of research institutions. One of our motivations has been to analyze what roles universities and faculties play in the resolution of the perceived "crises" in scholarly communication. (And there are of course a number of crises that are field-dependent.) Our premise is that disciplinary traditions and culture matter significantly in both predicting possible futures and the success or failure of policies that attempt to dictate scholarly behavior.

---

[1] Presentation slides are available at http://sites.nationalacademies.org/PGA/brdi/PGA_064019.
[2] The Future of SC Project Website and Associated Links:
Project site: http://cshe.berkeley.edu/research/scholarlycommunication.
Many of the arguments around sharing, time, and credit made here are given in more detail in Acord and Harley (in press), Harley et al. 2010, Harley and Acord, 2011).
[3] Disciplines included Anthropology, Biostatistics, Chemical Engineering, Law and Economics, English-language Literature, Astrophysics, Archaeology, Biology, Economics, History, Music, and Political Science. All the interviews have been published as part of thickly described disciplinary case studies. The entire research output is online and open access at http://escholarship.org/uc/cshe_fsc.
[4] Interview protocols covered a variety of broad questions: Tenure and promotion, making a name; Criteria for disseminating research at various stages (publication practices, new publication outlets, new genres); Sharing (what, with whom, when, why or why not?); Collaboration (with whom, when, why or why not?); Resources created and consumed: needs, discoverability, priorities, data creation and preservation; Public engagement; The future.

As I was exploring the issues related to this particular workshop and the various references the organizers had assembled for the meeting, I found this quote from the Australian National Data Service website of particular interest:

> Data citation refers to the practice of providing a reference to data in the same way as researchers routinely provide a bibliographic reference to printed resources. The need to cite data is starting to be recognised as one of the key practices underpinning the recognition of data as a primary research output rather than as a by-product of research. While data has often been shared in the past, it is seldom cited in the same way as a journal article or other publication might be. This culture is, however, gradually changing. **If datasets were cited, they would achieve a validity and significance within the cycle of activities associated with scholarly communications and recognition of scholarly effort.**[5]

The last statement (highlighted in bold) is actually quite complex and fraught. I will argue today why it has at least two questionable underlying assumptions. The first, is that by virtue of being citable, data achieve an equal footing with traditional publications in institutional merit review of scholars. The second, is that data standing alone, without an interpretive layer (such as an article or book) and without having been peer reviewed, will be weighted in tenure and promotion decisions the same as traditional publication.

The centrality of career advancement in a scholar's life

My argument (which I hope is not too circuitous) is that these two assumptions are contrary to what our research would suggest. As we have demonstrated (Harley et al., 2010), the primary drivers of faculty scholarly communication behavior in competitive institutions are career self-interest, advancing the field, and receiving credit and attribution. Although the institutional peer-review process allows flexibility for differences of discipline and scholarly product, a stellar record of high-impact peer reviewed publications continues to be the most important criterion for judging a successful scholar in tenure and promotion decisions. The formal process of converting research findings into academic discourse through publishing is the concrete way in which research enters into scholarly canons that record progress in a field. And, as the formal version "of record," peer-reviewed publication establishes proof of concept, precedence, and credit to scholars for their work and ideas in a way that can be formally tracked and cited by others. Accordingly, data sets, exhibitions, tools/instruments, and other 'subsidiary' products are awarded far less credit than standard publications *unless* they are themselves 'discussed' in a an interpretive peer-reviewed publication.

The importance placed by tenure and promotion committees, grant review committees, and scholars themselves, on publication in the top peer-reviewed outlets is growing, not decreasing, in competitive research universities (Harley et al., 2010: 7; Harley and Acord 2011). There is a concomitant pressure on all in the academy, including scholars at aspirant institutions globally, to model this singular focus on publish or perish, which we and others would argue translates into a growing glut of low-quality publications and publication outlets. This proliferation of outlets has placed a premium on separating prestige outlets (with their imprimatur as proxies for quality) from those that are viewed as less stringently refereed. Consequently, most scholars choose outlets to publish their work based on three factors: (1) prestige (perceptions of rigor in

---

[5] Australian National Data Service: http://www.ands.org.au/guides/data-citation-awareness.pdf.

peer review, selectivity, and "reputation"), (2) relative speed to publication, and (3) highest visibility within a target audience (Harley et al., 2010: 10). As we determined, this system is not likely to disappear soon and certainly will not be overturned by the adoption of new practices by young scholars, who hew, often slavishly, to the norms in their discipline in the interests of personal career advancement.

Securing credit and attribution

We note a continuum by field in how scholars receive attribution for their ideas.. In highly competitive fields like molecular biology, which have a race to publish and a fear of being scooped, early sharing of ideas, data, and working papers is almost unheard of. The archival journal article takes precedence and seals a scholar's claim on ideas. In relatively smaller (and often high paradigm and/or emerging) fields, with low internal competition, informal mechanisms for reputation management can enforce attribution because academic communities are centrally organized and maintained through face-to-face interaction (often via conferences and workshops). This sharing culture can change, however, with funding and other exigencies of a field. To give just one example, although economics is commonly described as a big sharing group where we are very open about ideas (and has a thriving environment of working papers, Harley et al., 2010: 357), the subfield of neuroeconomics is moving towards less sharing and mirrors practices in some of the biological sciences.

The importance of filters and managing time

As scholars prioritize their core research activities, they struggle to keep up to date and they look for more filters, not fewer, in determining what to pay attention to. In fact, time, and the related need for filters, is cited as one of the most influential variables in a scholar's decision whether to adopt new scholarly communication practices (Harley et al., 2007, 2010: 97). Most scholars turn to the familiar filters of peer review, perceived selectivity, reputation, and personal networks to filter what they pay attention to.

We would argue that, given this background, and an exceptionally heavy reliance on peer review publications to aid the tenure and promotion committees, it comes as no surprise, that competitive scholars divert much of their energies and activities toward the development and production of archival publications and the circulation of the ideas contained within them rather than focusing, for example, on curating and citing data sets.

Variation in how different scholarly outputs are weighed

Lest you think the way in which scholarship is credited in tenure and promotion decisions in research universities is binary, it is important to note that teaching, service, and the creation of non-peer reviewed scholarship such as data bases or tools, are most certainly credited, but they do not receive as much emphasis as peer reviewed articles or books published in prestigious outlets. Some things are weighted more heavily than others (and what is weighted is field-dependent). For example, people can get credit for developing software or an important database, but that would rarely be the sole criteria in most fields and not given equal weight as publications. We heard again and again that new genres of scholarship in a field are acceptable as long as they are *peer reviewed*. Examples abound on the emphasis in tenure and promotion based on *interpretive work* versus cataloging or curating. Data curation and preservation alone are simply not considered to be a high level of scholarship.

Let me give you some examples. In biology, new and emerging forms of scholarship are not valued much in their own right, but supplemental articles describing a novel tool or a database may be considered positively in review. "Just" developing software or data resources can be perceived as less valuable tool development, rather than scholarship. In history, a scholar who published only footnoted sources, but no interpretation of the sources in the form of a peer reviewed book published by a prestigious press, would not be promoted. That is, new and emerging forms of scholarship (e.g., curating data sets, creating Web-based resources, blogs) are valued only insofar as they are ancillary to the book/monograph.

In political science, scholars often create and publish datasets. Similar to the biological sciences, these efforts can earn a scholar increased visibility when other researchers use their data. Significant institutional credit is only received for this work, however, if a strong peer reviewed publication record, based on the dataset, accompanies it. In archaeology, developing and maintaining databases or resource websites, which is common, is considered a research technique or a service to scholarship but not a substitute for the monograph. In astrophysics (and despite the reliance on arXiv for circulation of early drafts, data, and so on), developing astronomical instrumentation, software, posting announcements, and database creation are considered support roles and are usually ascribed a lower value in advancement decisions than peer reviewed publications.

How will datasets be peer reviewed?

What, do you ask, does any of this discussion have to with motivating scholars to abandon their traditional data creation, citation, and sharing practices in the face of calls (and sometimes mandates) from some journals and funding bodies to publish data sets, particularly in the sciences and quantitative social sciences? These are powerful calls that are motivated by the desire for more transparency in research practice, greater returns on funders' investments, as well as claims that the growing availability of digital primary source material is creating novel opportunities for research that is significantly different than traditional forms of scholarship. We predict, however, that despite this power, changes in data management practices, as with in-progress scholarly communication, will be heavily influenced by matters of time, credit, personality, and discipline.

I would suggest that the most important question for motivating scholars to conform fully to developing new practices is, How will data creation and curation ever be weighted similarly to traditional publications if not peer reviewed? And who will do the peer review and how? I cannot emphasize enough that we just do not know how or when data will be formally peer reviewed in the same way that journals and books are currently.

Some presume that "open" peer review, a free-for-all, crowd-sourced system, will solve this problem.[6] I would reply that such a system would be loaded with intractable problems, not least of which is that scholars, perhaps especially senior scholars, already spend an enormous amount of their time conducting peer review in its myriad forms: evaluating grants, writing letters of reference, mentoring graduate students, responding to emails for feedback on work, and so on. The result is that even established publishers have an exceptionally difficult time recruiting competent reviewers (Harley and Acord, 2011: 25), and most scholars find it difficult to spare the time to conduct these formal reviews, let alone engage in "optional" volunteer and open reviews that will not likely contribute to career advancement.

It is our opinion (which we review at length in Harley and Acord, 2011: 45-48 and Acord and Harley 2011), the lack of uptake of open peer review in a variety of experiments, where commentary is openly solicited and shared by random readers, colleagues, and sometimes editor-invited reviewers (rather than exclusively organized by editors), is not likely to be embraced by the academic community anytime soon. The results of only a few of these experiments indicate that open peer review *might* have the potential to add value to the traditional closed peer-review process, *but* that it also exacts large (probably unsustainable) costs in terms of editor, author, and reviewer time. Scholars are likely to avoid *en masse* such experiments because many do not have the *time* to sort through existing publisher-vetted material, let alone additional "unvetted" material or material vetted by unknown individuals. In sum, open peer review simply adds one more time consuming circle of activity to a scholar's limited time budget. Telling support is provided by the recent policy shift of *The Journal of Neuroscience* (Maunsell, 2010) and the *Journal of Experimental Medicine* (Borowski, 2011), which recently announced their decisions to cease the publication of supplementary data because reviewers cannot realistically spend the time necessary to review that material closely, and critical information on data or methods needed by readers can be lost in a giant, time-consuming "data dump." An editorial in the *Journal of Experimental Medicine*, titled "Enough is Enough"[7] makes the case:

> Complaints about the overabundance of supplementary information in primary research articles have increased in decibel and frequency in the past several years and are now at cacophonous levels. Reviewers and editors warn that they do not have time to
>
> scrutinize it. Authors contend that the effort and money needed to produce it exceeds that reasonably spent on a single publication. How often readers actually look at supplemental information is unclear, and most journal websites offer the supplement as an optional (item to) download…

---

[6] Or, many will argue that alternative "bibliometrics" measuring popularity and use is the solution. A consequence of the 'inflationary currency' in scholarly communication is a growing reliance on bibliometrics, such as the impact factor, and an increasing 'arms race' among scholars to publish in the highest impact outlets. As detailed by Harley and Acord (2011: 48-53), there is widespread concern that at this time, and taken alone, alternative (quantitative) metrics for judging scholarly work are much more susceptible to gaming and popularity contests than traditional peer-review processes.

[7] Enough is Enough" Christine Borowski, July 4, 2011 *Editorial* published in the *Journal of Experimental Medicine* (*JEM*).

In sum, data sharing is greatly impeded by scholars' lack of personal time to prepare the data and necessary metadata for deposit and reuse (which includes the sometimes Herculean efforts of converting analog data to digital formats, migrating old digital formats to new ones, or standardizing messy data). For scholars focused on personal credit and career advancement, narrowly defined, there is no advantage to spending time (and grant funding) curating or peer reviewing data, when that same time can be applied to garnering support for the next research project and/or publishing and peer reviewing books and articles. While data sharing may be facilitated by development of new tools and instruments that ensure standardization (such as in gene sequencing), the idiosyncratic ways in which scholars work, and the extreme heterogeneity of data types in most non-computational fields, do not lend themselves to one-size-fits-all models of data sharing. The escalation of funder requirements (e.g., NSF, NIH) for sharing data management plans points to an important space for social scientists to track. We, and others, predict that faculty will not be doing the actual work, but rather a new professional class and academic track (perhaps akin to museum curators, specialist librarians, or tool-builders) may emerge to take on these new scholarly roles (cf: Borgman, 2007; Nature, 2008; Science, 2011; Waters, 2004). They of course will need to be paid and regularized in some fashion. In sum, until issues of time, credit, and peer review are worked out, we predict an uneven and slow adoption by scholars of sharing, curating, and publishing data openly, and hence the citation and attribution of same.

References

Acord, S.K. and D. Harley (in press). "Credit, Time, and Personality", invited by *New Media and Society*. Available for open peer review at nms-theme.ehumanities.nl/manuscript/credit-time-and-personality-acord-and-harley

Borgman, C.L. (2007) Scholarship in the Digital Age: Information, Infrastructure, and the Internet Cambridge, MA: The MIT Press.

Borgman, C.L .(2011) The conundrum of sharing research data. *Journal of the American Society for Information Science and Technology*. (accessed 19 October 2011)

Borowski, C. (2011) Enough is enough. *Journal of Experimental Medicine* 208(7): 1337.

Harley, D. and Acord, S.K. (2011) *Peer Review in Academic Promotion and Publishing: Its Meaning, Locus, and Future*. University of California, Berkeley: Center for Studies in Higher Education. Available at: http://escholarship.org/uc/item/1xv148c8.

Harley, D., Acord, S.K., Earl-Novell S., Lawrence, S., and King, C.J. (2010) *Assessing the Future Landscape of Scholarly Communication: An Exploration of Faculty Values and Needs in Seven Disciplines*, University of California, Berkeley: Center for Studies in Higher Education. Available at: http://escholarship.org/uc/cshe_fsc.

Maunsell, J. (2010) Announcement regarding supplemental material. *The Journal of Neuroscience* 30(32): 10599-10600.

Nature (2008) Special Issue: Big Data. *Nature* 455(7209).

Science (2011) Special Online Collection: Dealing with Data. *Science* 331(6018). (Accessed 19 October 2011) http://www.sciencemag.org/site/special/data/.

Waters, D.J. (2004) Building on success, forging new ground: The question of sustainability. *First Monday* 9(5).

DISCUSSION BY WORKSHOP PARTICIPANTS

Moderated by Paul F. Uhlir

PARTICIPANT: I was pleased that Sarah Callaghan mentioned the European database directive. What is the status of the database protection legislation in the United States? In the House of Representatives there were the Moorehead Bill in 1996, and the Coble Bill in 1997 and again in 1999, and in the Senate, the Hatch bill in 1998. Then such legislation just disappeared. What happened?

MR. UHLIR: From 1996, there was a coalition of internet service providers, the universities, the libraries, and the academics opposed to the database protection bills. They were just barely hanging on in terms of keeping the legislation from becoming enacted. In 2001, I believe, the Chamber of Commerce made an assessment of the costs and benefits of the law and, since they are not primarily database providers but users, they determined that it would raise the cost of doing business. They said they would keep track of every vote in favor and keep that in mind when they dole out the re-election money. So, that was the end of those legislative proposals and I have not heard about any new proposal since.

PARTICIPANT: I want to ask a question about credit and why anyone would want to share their data. There are real dangers involved in sharing data, not just the work. I thought I would raise it and let someone expound.

DR. HARLEY: Not all data are created equal and not all data want to be shared. I think Christine Borgman has done good research of the reasons why data are not shared.

We think personality has a lot to do with it and that is why some people are sharing and some people are not. It has to do with preprints and early writings as well. When you submit a paper and it is accepted, you have to have some kind of data management plan. Even when journals require data to be available and to be shared, however, we found that less than 10 percent of requests were honored for sharing data. It all goes back to the attitude of: "when I am finished with it, I am happy to share it." In archeology, for example, that can go on for twenty years.

MS. SMITH: I agree. I think there are strong disciplinary differences in attitudes related to the sharing of data. Some fields have the habit of sharing more than others. There are certainly some fields where sharing is not even discussed. The other issue is that even if it is something researchers are willing to do, they are not going to do it if they have to spend time on it. So, if we made it easy, they might be more inclined to do it.

PARTICIPANT: I have asked investigators from biology if they think their data will be cited if they are forced by journal publishing policies to make them available. Most of them said yes, they thought that they would be cited. My next question was: do you think that citation will be valued by your institutions in peer review of your work? The answer was no.

MS. SMITH: My comment is not based on any actual research, but just my impression from talking to a lot of people. They are sharing their data just because it is the right thing to do. When you can share and there are no negative consequences, you will do it as long as you do not have to spend a huge amount of time on it. I see this culture becoming more common and policies like those of the NSF are helping. Maybe it will just happen naturally and that we are overemphasizing citation as an incentive. It may be that they just want easier ways to share data because they realize it is the right thing to do.

DR. CHAVAN: As I keep going through all these presentations and especially in this particular session, it becomes clearer to me what the challenges for data citations are. Technology can certainly help in moving in the right direction, but I think that the mindset of those who publish the data and those who are involved in the data management life cycle itself is very critical in moving forward. It is a social and cultural consideration and there is no one single solution to that. I think it has to be worked on every level of the data life cycle.

MS. SMITH: I do not disagree with that, but I think the technology does have a big effect on the costs of this process, and social practices depend a lot on costs.

PARTICIPANT: How much of the data that we care about to drive science is not generated by people who are academic faculty looking for tenure and promotion? Coming from a Federally Funded Research and Development Center (FFRDC), the intellectual property issues are drastically different than those at a university. What are the social factors that affect those generating data in other types of institutions? Those people also want credit.

MS. SMITH: There are many other players in this ecosystem, such as libraries, but they do not depend on citation that much, which I think is why we are not talking about it here. There are different reward mechanisms for librarians and for data curators in national archives.

PARTICIPANT: Do you have any examples of people saying "we have these citations norms, but we do not want to enforce them?" Are there societies with articulated policies that are not legally binding so that they can get around the notion of enforcing credit via the law and instead, have clearer norms?

MS. SMITH: That is how citation works now. I am not legally obliged to cite your article when I use your ideas. There is no intellectual property law that requires me to do that.

DR. GROTH: So then why do we need to discuss these issues?

MS. SMITH: There nonetheless are a lot of researchers using contracts, Creative Commons (CC) licenses, and data usage license agreements who want to be able to mandate credit.

MR. PARSONS: Just a couple of comments on the issue of open data citation as an incentive. I spent too much of my life working on a big international project called the "International Polar Year," and it had what I think was a landmark data policy pushing for not only open data but also timely release of data. This caused quite a bit of controversy in a lot of different disciplines. We developed citation norms and there is a document associated with the Polar Information Commons, where we have actually documented some of these norms, but that is not what motivates people to share. I think there are two things that motivate people to share. Every time I share data, I learn something and I think that is partly the personality aspect that was mentioned earlier. Sharing is also moving the field forward. So if people can share in a way that makes them feel that they are collaborators, they will be happy to share. However, the most effective way in getting people to share was when the funding agencies required it by saying that researchers will not get their next year's funding or the next grant if their data are not available.

DR. HARLEY: I think you are right. I think the stick is probably your best tool, given the way people react, and given their personal motivations, concerns, and fears.

MS. SMITH: If there is a requirement from an agency to share the data in order to get the next grant, researchers will need to prove to the agency that they met the requirement. So, citation does come back as maybe a simpler way of proving that the data were actually shared, because the data have an URI or are deposited at a reputable data archive.

MR. WILBANKS: The main reason why we are talking about these issues is that there is a growing push from the open science movement to put CC waivers or licenses on data, and to draft other licenses for data and databases.

DR. MINSTER: This notion that data work is good science is just fine, but does not obviate the need to address other issues that Diane talked about. We have to change the mindset of those engaged in the scientific process and give proper recognition and advancement to people who spend their life doing this, if they do it well.

DR. HARLEY: I think it starts with the investigators mentoring of their graduate students and post docs. It is the senior scientists setting an example of good practice for those younger individuals in their labs and projects. For example, I know a researcher at UCSF who makes very clear his guidelines with regard to good scholarship in his lab: what he will and will not publish, what are good and bad practices. I also want to emphasize that it is not a black and white issue. It should be acceptable to have different gradations of credit for different cases.

DR. HELLY: The issue about carrots and sticks is kind of a false dichotomy in some sense. The sticks are helpful in encouraging proper data practices and they indicate that the agency values and gives credit for delivering the data products. However, it does not in any way do enough to get the data products to a quality great enough to ensure that other people can actually use those data constructively. People will just take their data and dump them in these repositories. The real motivation that was discussed here is the basic scientific realization that there has to be good practice and that this practice has to be learned and taught. This has not been done. Once we get to that stage, the community minded people would adhere to these policies, especially in the sciences where community data resources are a very powerful tool for people doing individual science. I think in the earth sciences, in particular, this is a very strong motivation, where there are global issues and data have to be put together across the globe.

DR. BORGMAN: I just want to add two policy issues that I am surprised have not come up yet, and I would like the panel to address them. One is the role of embargoes, which are in place in many fields. Employees protect the investigators and the funding agencies. Those who impose embargo periods generally pick a time period that is long enough for people to get their publications out, but short enough to encourage getting the data out. The other point is about data registries. We have found in our research that even if people are not willing to deposit their data, they are willing to register their data with metadata and then one can at least find them. The pointer might be a telephone number or an URI, but the data registry turns out to be a lower bar.

PARTICIPANT: To the extent embargoes are imposed from outside or internally depends on the discipline itself.

MS. SMITH: Yes, embargoes are very important to people. I agree that is key, but the registry is an interesting point because I actually cannot remember a case in which that was an option for people. If registries exist, they are new. I know there are efforts to build them, but it is not on the radar of most of the researchers I have talked to. They think that sharing means putting their data on the web or giving them to an archivist. That might change the culture a little bit, but we have to have better examples of it that people can see.

PARTICIPANT: I am just wondering if in any of your studies whether you have found any correlation between how much data sharing there is versus how much money is in the system? Because the funding levels fluctuate over the years, my intuition would tell me that people are less likely to share when there is less money.

MS. SMITH: This is a very interesting question, I would have guessed the same thing, but that is not what I see in practice, in the health sciences for example, where there is a lot of money. In practice, we are seeing more data sharing in the health sciences than other fields where there is less funding.

DR. HARLEY: One of the things we talked to people about in our research is what do you see in the future? There appear to be different trajectories, depending on the history and economics of the discipline. They can be rather divergent. Whether or not the value systems are going to change with the way data are captured and described in these fields, is unknown at this point.

PARTICIPANT: There is an important question here about data citation and whether that is what we are asking people to do? I think we should keep in mind that data publication and data citation are both metaphors taken from the journal publication system, and I think metaphors are tricky. The danger with any metaphor is to say that it is exactly the same. We want to define what is it that we want to consider about data and how we want the data to be used differently from journals, but still using that same sort of metaphor.

PART FOUR

EXAMPLES OF DATA CITATION INITIATIVES

13- The DataCite Consortium

Jan Brase[1]
DataCite

I would like to give you an overview of what DataCite is doing. The idea behind our project is that science is global and therefore we need global standards to do it. We need some global workflows and cooperation between global players such as global data centers and publishers. Of course, science is also carried out locally. Scientists themselves do global science but they are embedded in their local institutions, libraries, and funding agencies. So, that is the paradox that we are dealing with in our project. We want to have global answers but at the same time want to act on a very local basis. That was the main motive behind founding DataCite as a global consortium of local institutions.

On the one hand, DataCite is carried out by members like the Technische Informations Bibliothek (the German Library of Science and Technology), the California Digital Library, and the British Library. These institutions act locally with local data centers. For example, the British data center works with the British Library and uses their services as their local coverage partner. On the other hand, DataCite itself as a global organization consisting of other global organizations, such as publishers. This is important, because the publishers now only have one central partner to work with on data citations and do not have to deal with the dozens of data centers individually on the local level.

This is the list of the current membership of DataCite. There are the 15 members from 10 countries.

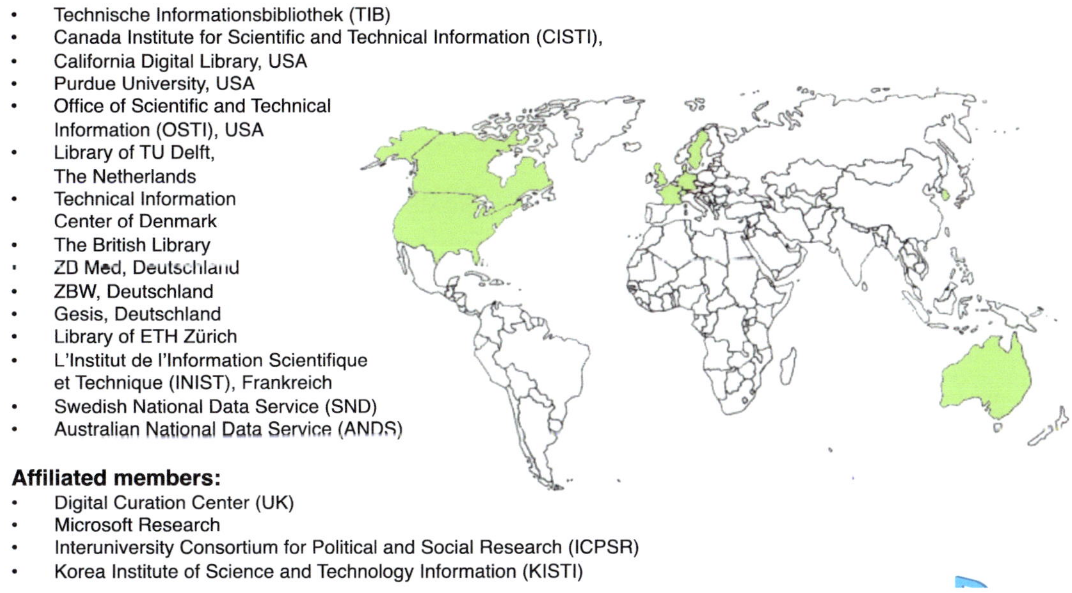

- Technische Informationsbibliothek (TIB)
- Canada Institute for Scientific and Technical Information (CISTI),
- California Digital Library, USA
- Purdue University, USA
- Office of Scientific and Technical Information (OSTI), USA
- Library of TU Delft, The Netherlands
- Technical Information Center of Denmark
- The British Library
- ZD Med, Deutschland
- ZBW, Deutschland
- Gesis, Deutschland
- Library of ETH Zürich
- L'Institut de l'Information Scientifique et Technique (INIST), Frankreich
- Swedish National Data Service (SND)
- Australian National Data Service (ANDS)

Affiliated members:
- Digital Curation Center (UK)
- Microsoft Research
- Interuniversity Consortium for Political and Social Research (ICPSR)
- Korea Institute of Science and Technology Information (KISTI)

FIGURE 13-1 List of DataCite members.

---

[1] Presentation slides are available at http://sites.nationalacademies.org/PGA/brdi/PGA_064019.

The next question is what are we doing? Simply, DataCite is by definition a registration agency of Digital Object Identifiers (DOIs). DataCite assigns DOI names to datasets. This is one of our main pillars. We are also actively involved with our members to work on standard definitions. We try to activate those standards for data citation. We also have plans to establish a central metadata portal, where you will have free access to the metadata of all content that we have registered.

One of our main functionalities is to provide DOI names for datasets. We all agree that identifiers for datasets are important to make data citation possible. Let me explain with an example of citations for a dataset:

Storz, D. et al. (2009):
Planktic foraminiferal flux and faunal composition of sediment trap L1_K276 in the northeastern Atlantic.
http://dx.doi.org/10.1594/PANGAEA.724325.

As a supplement to the article:
Storz, David; Schulz, Hartmut; Waniek, Joanna J; Schulz-Bull, Detlef; Kucera, Michal (2009):
Seasonal and interannual variability of the planktic foraminiferal flux in the vicinity of the Azores Current.
Deep-Sea Research Part I-Oceanographic Research Papers, 56(1), 107-124,
http://dx.doi.org/10.1016/j.dsr.2008.08.009.

First, we have a dataset citation. You will see that one of the good things about using DOI names is that it actually has the same look and feel as a classical journal citation, with the title and the DOI name that you can click on to access the article. It also has the data center as the affiliation. The most useful part, however, is that you can click on the DOI to directly access the data. This allows you to download the data into your system for your own analysis or visualization. If you decide to reuse the data for your own work, the fact that the data has a DOI name allows you to cite the data and give the original author credit for using those data.

If the user goes to the webpage of the article, that person then sees that there is data available for this article. In contrast to the article that may only be available to paying customers (subscribers), the access to the data is free of charge. So, if the user is interested in the article, he or she can look at the data first and then decide what to do.

That is one of the fundamentals of DataCite—we believe that the data that support the article should be freely available. In cooperation with the publishers, we have designed our system in a way that even if you do not have the right to look at the article (without paying) and you can only access the abstract and the table of contents, the link to the data is displayed and access to the data is free of charge. In a way, this is a win-win situation for the publishers because the availability of the data enhances the value of the article and promotes its use, while the publishers do not any lose revenues.

Finally, let me briefly tell you where are we at this point. DataCite has registered over a million records with DOI names. We have also published metadata schema[2] that we use for all records. We just released the Beta Version of DataCite Metadata Store online in July 2011: see http://search.datacite.org.

In 2012, we expect to have around 800,000 datasets in the Metadata Store and hope to have all records available in the middle of the year. As for next steps, we are working with other organizations, such as Thomson Reuters Science, to index our content. The metadata are freely available to harvesters via http://oai.datacite.org. We are also working with Elsevier and the Pangaea data center and other publishers to find more article-data links.

---

[2] http://schema.datacite.org.

14- Data Citation in the Dataverse Network®

Micah Altman[1]
Director of Research, MIT Libraries
Head/Scientist, Program for Information Sciences
Non-Resident Senior Fellow, The Brookings Institution

Overview of the Dataverse Network

The Dataverse Network ® (DVN) project is an open source application for discovering, publishing, citing, and preserving research data. (See Crosas 2011 for a detailed description) Created by Gary King (2007), and based on over a decade of digital library research (Altman, et al. 2001), the DVN has been described as the "state of the practice" in open source data sharing infrastructure (Novak, et al. 2011).

The DVN functions as an institutional repository system designed specializing in quantitative data, a federated library and catalog system, and as virtual archiving system. There are multiple installations of the DVN software. Each "Dataverse Network" is hosted on its own host server, and hosts numerous virtual archives, each known as a "dataverse." A Dataverse can act as a virtual archive for a journal, a department, a research group, an individual, or a library. The DVN server allows an institution to provide unified backup services, citation generation, discovery, preservation and other sorts of core services while enabling the owners of the virtual archives to manage content deposit, dissemination, terms of use, and branding.

Dataverse Networks can be used to host metadata, text, images, and data from many different disciplines. The first version of the DVN allowed deposit of any format, and provided enhanced services for individual microdata tables. Support for other data structures, such as social network data and time-series, have been added as the software evolved and as the evidence base of social science research has shifted (see Altman & Klass 2005; Altman & Rogerson 2008).

The first installation of the DVN was at the Harvard Institute for Quantitative Social Sciences (IQSS). That installation has been officially incorporated into the Harvard Library system, but remains open to researchers world-wide. It provides search, data access, and data hosting, at no charge. The Harvard DVN functions also as a federated catalog of over 50,000 social science data data sets, including the holdings of the Data Preservation Alliance for the Social Sciences [Altman, et al. 2009] and the Council of European Social Science Data Archives. The current holdings of the Harvard DVN are predominantly used for social science research.

Additional Dataverse Networks are hosted at the University of North Carolina, in Thailand, and elsewhere around the world. Most DVN's are federated using the open-archives metadata protocol, and so almost five hundred virtual archives may be searched and browsed from a single location.

---

[1] This is an updated, corrected, and expanded version of the original presentation, which is available at http://sites.nationalacademies.org/PGA/brdi/PGA_064019.

A data citation in depth

We will start with a brief example. Figure 14-1 below shows the home page from my own scholarly virtual archive (or "dataverse"). This is virtual archive contains all of the data that I have disseminated as part of my own research. Note that although this virtual archive is organized as a single collection, many of the datasets listed actually are distributed by different archives, and may be stored in different locations.

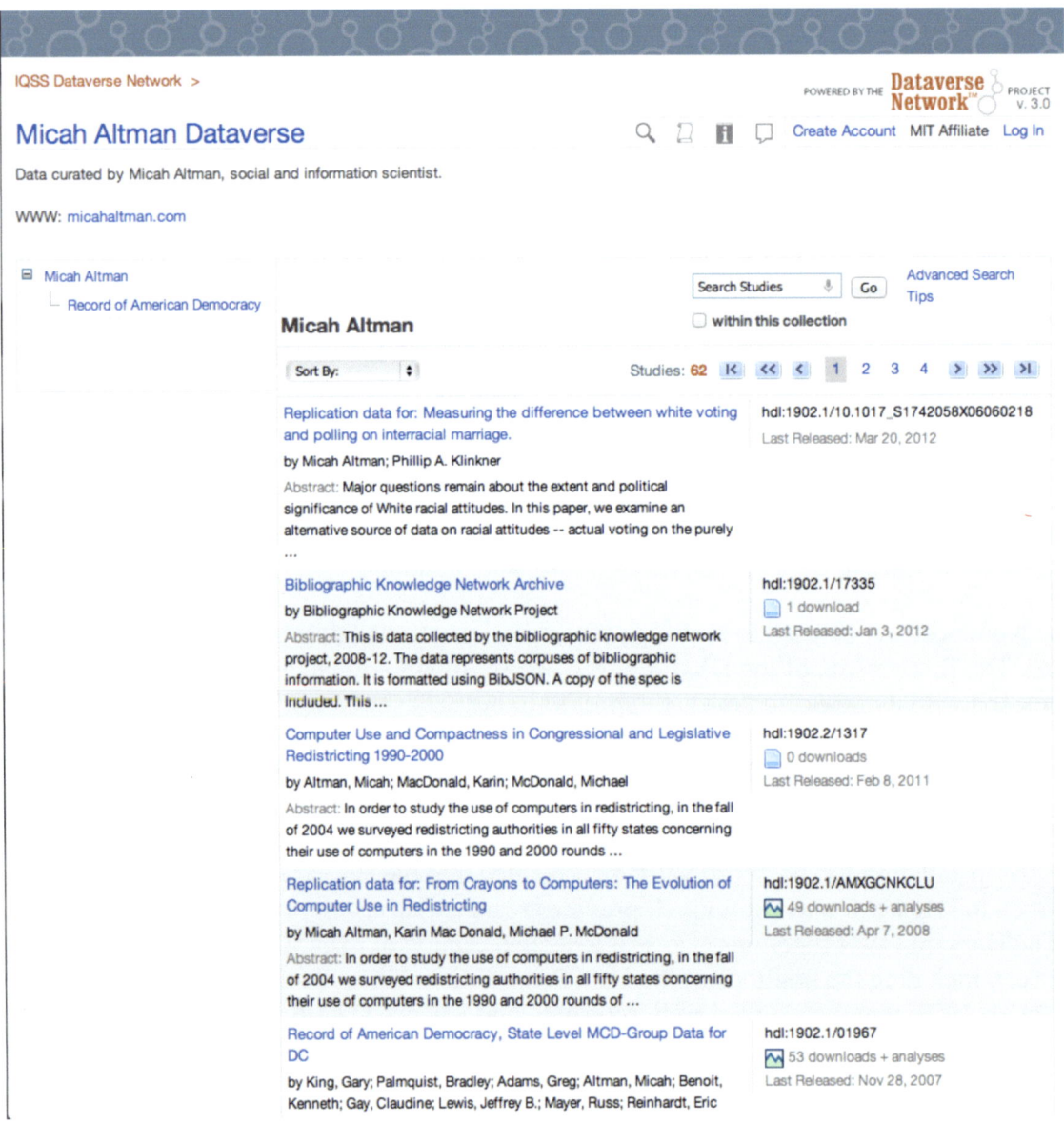

FIGURE 14-1 Micah Altman's Dataverse.

Shown below is the catalog page for a particular dataset in this collection. The catalog page for this dataset shows descriptive information about it. The "data and analysis" tab (not shown), allows users to download the data itself, and to perform on-line analyses.

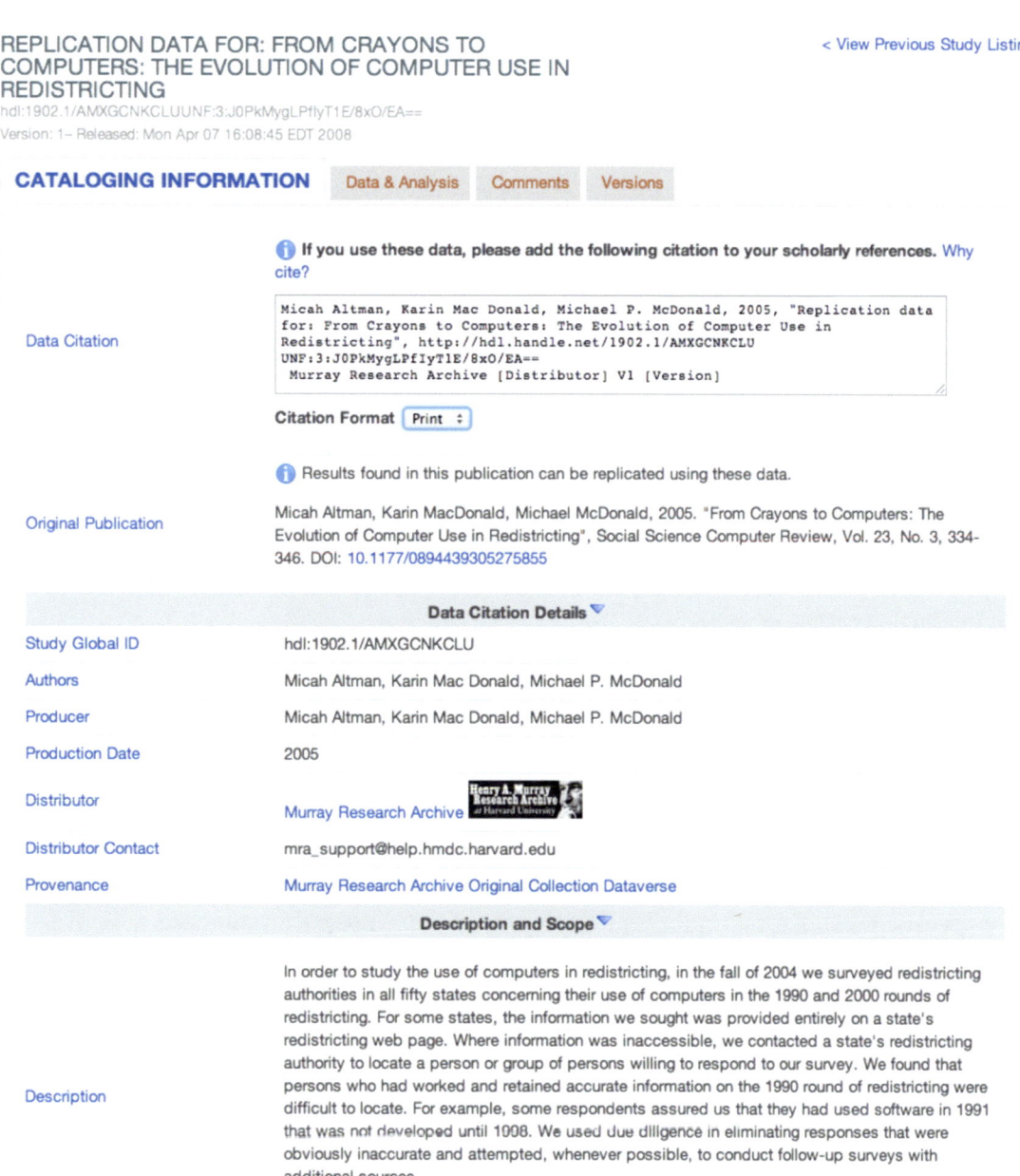

FIGURE 14-2 Catalog page for data to replicate "From Crayons to Computers: The Evolution of Computer Use in Redistricting."

While this dataset is in my own scholarly dataverse, it is formally "distributed" by the Murray Research Archive, as indicated through the distributor information in the citation above. This indicates that the dataset was reviewed by the Murray Archive and complied with its selection policies. This also indicates that the Archive takes "archival" responsibility for it, and will continue to ensure its long-term availability. Other datasets in my collection are distributed by ICPSR, and by the IQSS Dataverse Network.

The DVN system provides several varieties of citation related to the dataset. One citation, shown below, is used for articles related to the dataset. In this case, the dataset is explicitly identified as a "replication dataset", containing all of the information necessary to replicate a published article.

> ⓘ Results found in this publication can be replicated using these data.
>
> Original Publication — Micah Altman, Karin MacDonald, Michael McDonald, 2005. "From Crayons to Computers: The Evolution of Computer Use in Redistricting", Social Science Computer Review, Vol. 23, No. 3, 334-346. DOI: 10.1177/0894439305275855

FIGURE 14-3 Citation for original publication with which data are associated.

The printed citation for the dataset is shown below. Note that it shows a separate persistent identifier and separate publication information, because, in this case, data and publication are distributed separately.

> Data Citation
>
> ```
> Micah Altman, Karin Mac Donald, Michael P. McDonald, 2005, "Replication data for: From
> Crayons to Computers: The Evolution of Computer Use in Redistricting",
> http://hdl.handle.net/1902.1/AMXGCNKCLU UNF:3:J0PkMygLPfIyT1E/8xO/EA==
> Murray Research Archive [Distributor] V1 [Version]
> ```

FIGURE 14-4 Printed data citation example.

The Dataverse Network software produces citations that are based on a community standard first proposed by Altman and King (2007). The hundreds of virtual archives ("dataverses") using the system automatically produce these citations for all of their datasets. This standard is also in use by the member archives of the Data Preservation Alliance for Social Sciences (Data-PASS 2012).

The Altman-King standard is light-weight, as it requires a minimal number of elements, and does not impose a specific presentation on these. Data citations that follow this standard will, at a minimum, include the following information: *author(s)*, *date*, *title*, and a *persistent identifier* that is drawn from any standardized and web-resolvable identifier scheme (e.g., DOI, handle, PURL, LSID). [2]

This minimal citation standard has a number of extensions:

1. First, where feasible, the standard recommends the inclusion of a URI corresponding to the persistent identifier; a Universal Numeric Fingerprint, which provides fixity information; and explicit versioning information.
2. Second, a default ordering of the elements is suggested for presentation and parsing—but elements may be reordered if explicitly labeled.
3. Third, the standard is extensible, and citations can include elements from any other XML-ized citation schema, simply by using XPath syntax and labeling conventions. The Dataverse Network System currently uses these extensions to indicate the producer and distributor for the dataset.

---

[2] In addition, Data-PASS recommends that (a) cited data be deposited in a archive, and (b) data citations be included in along with other publications in journal articles and indices—generally data citations should not be presented only in an ad-hoc location, such as the publication text, acknowledgement, figure labels or substantive footnotes.

4. Fourth, the standard provides an extension for citing subsets (portions) of the dataset. The Dataverse Network System uses this to create citations for each of the data extracts it produces.

These points are illustrated in the citation below, which is associated and drawn from a separate study in the independent Odum Dataverse Network. This citation includes optional extension fields imported from the Data Documentation Initiative (DDI) metadata standard, which are used to indicate the data distributor, version, and the attributes of the data subset extracted. Note that separate UNF's are calculated for both the subset extracted and the dataset in its entirety.

> Center for Survey Research, 2011-12-08, "Firearms 2004,"
>
> http://hdl.handle.net/1902.29/10863 UNF:5:kkz60UZUqLvi8cKT3i82GA==
>
> Odum Institute for Research in Social Science [Distributor] V1 [Version]
>
> CASEID,COMMUNIT,PEOPLE [VarGrp/@var(DDI)];
>
> UNF:5:WjxtjId/Gi9ICzCvBEEyiQ==

FIGURE 14-5 Data subset citation example.

The Universal Numeric Fingerprint (UNF) is a novel part of the Altman-King citation standard. (The first version of the UNF algorithm was developed in Altman, *et al* 2003, and it has been extended in Altman-King 2007, and Altman 2008.) A UNF is a short, fixed-length string that summarizes the entire dataset. Thus it provides "fixity," enabling a future user of the dataset to verify that the dataset they possess is semantically identical to the one originally cited in a publication. Similar to a cryptographic hash function, the UNF is tamper-proof, and will change if the values represented by the dataset change. Unlike a cryptographic hash, the UNF is invariant to the specific file format used to store/serialize the dataset. A UNF works by first translating the data into a canonical form; computing an approximation of that canonical form using a specified precision; and applying a cryptographic hash function to the approximation of the canonical object. The advantage of canonicalization is that it renders UNFs format-independent: if the data values stay the same, the UNF stays the same—even when the data set is moved between software programs, file storage systems, compression schemes, operating systems, or hardware platforms. Extensions to the UNF algorithm (Altman 2008), describe methods for computing UNFs over more complex objects, using recursion.

Data citation use cases and principles

Altman and King (2007) motivate their data citation standard on the basis of provenance, replicability, and attribution. Data citations should contain information that is sufficient to: locate the dataset that is referred to as evidence for the claims made in the citing publication; verify that the dataset one has located is the semantically identical to the dataset used by the original authors of the citing publication; and correctly attribute the dataset.

In the last five years, citation practices have evolved. To explore this issue, the Institute for Quantitative Social Science convened a workshop for leaders in publishing, data archiving, and data citation research, with the aims of identifying use cases and principles for data citation. The discussions facilitated through this workshop lead to a consensus on a number of additional core use cases, operational requirements, and principles.

Time prohibits describing these use cases and operational requirements in detail. To summarize they were grouped into five categories:

- Attribution. Data citations are used for enabling appropriate legal *and* scholarly attribution for the cited work.[3]
- Persistence. Citations are used to refer to, and to manage, objects that are persistent.
- Access. Citations are used to facilitate short and long term access to the object, by humans and by machine clients.
- Discovery. Citations are used to locate instances of the dataset; and as part of the process of discovering derivative, parent, and related works
- Provenance. Citations are used to associate published claims with evidence supporting them, and to verify that the evidence has not been altered.

Moreover, the workshop identified a "first principle" for citing data:
data citations should be treated as first-class objects for publication.

This principle has a variety of implications that depend on the specific context of publication. Notwithstanding, workshop participants identified two broad corollaries: Citations to data should be presented along with citations to other works—typically in a "references" section; and data should be made as easy to cite as other works—publishers should not impose additional requirements for citing data, nor should they accept citations to data that do not meet the core requirements for citing other works.

The workshop articulated three additional principles, based on discussions of the core uses of data citation:

- At a minimum, all data necessary to understand assess extend conclusions in scholarly work should be cited.
- Citations should persist and enable access to fixed version of data at least as long as the citing work exists.
- Data citation should support unambiguous attribution of credit to all contributors, possibly through the citation ecosystem.

These principles have implications for the entire ecosystem of publication—there are implications for authors, editors, publishers, and software developers. Neither standards documents nor software can ensure that data is cited properly—but both can help. The Dataverse Network System, and the citations produced through it, appear consistent with these principles.

---

[3] Note that these scholarly and legal attribution are conceptually separable: the first type of attribution is defined in terms intellectual property rights, whereas the second is defined in terms of scholarly norms.

References

Altman, M., Gill, J., & McDonald, M. (2003). Numerical issues in statistical computing for the social scientist. New York: John Wiley & Sons.

Altman, M. (2008). A Fingerprint Method for Scientific Data Verification. In T. Sobh (Ed.), *Proceedings of the International Conference on Systems Computing Sciences and Software Engineering 2007* (311-316). New York: Springer Netherlands. Retrieved from http://www.box.net/shared/0x8ld06hceg0ltpjyfu4

Altman, M., & Crabtree, J. (2011). Using the SafeArchive System : TRAC-Based Auditing of LOCKSS. *Archiving 2011* (165-170). Society for Imaging Science and Technology. Retrieved from http://www.imaging.org/IST/store/epub.cfm?abstrid=44591 http://www.box.net/shared/8py6vl9kxivo6u21rkn8

Altman, M., and King, G. (2007). A Proposed Standard for the Scholarly Citation of Quantitative Data. *DLib Magazine*, *13*(3/4), 1082–9873. SSRN. Retrieved from http://www.dlib.org/dlib/march07/altman/03altman.html.

Altman, M., and Klass, G. M. (2005). Current research in voting, elections, and technology. *Social Science Computer Review*, *23*(3), 269-273. Retrieved from http://ssc.sagepub.com/content/23/3/269.full.pdf. http://www.box.net/shared/1xlf6n7erk2bgu34z0nj.

Altman, M., Andreev, L., Diggory, M., King, G., Sone, A., Verba, S., Kiskis, D. L., et al. (2001). A digital library for the dissemination and replication of quantitative social science research: the Virtual Data Center. *Social Science Computer Review*, *19*(4), 458-470. Sage Publications. Retrieved from http://www.box.net/shared/d3cf8u0gtyml2nqq3u2f.

Altman, M., Crabtree, J., Donakowski, D., Maynard, M., Pienta, A., & Young, C. (2009). Digital Preservation Through Archival Collaboration: The Data Preservation Alliance for the Social Sciences. *The American Archivist*, *72*(1), 170-184. Retrieved from http://archivists.metapress.com/content/EU7252LHNRP7H188.

Altman, M. and Rogerson, K. (2008). Open Research Questions on Information and Technology in Global and Domestic Policis - Beyond "E-." PS: Political Science and Politics XLI.4 (October, 2008): 835-837. Retrieved from: http://journals.cambridge.org/action/displayFulltext?type=1&fid=2315612&jid=PSC&volumeId=41&issueId=04&aid=2315604 http://www.box.net/shared/2tlvqxgnzu5p5sby2mzq.

Data Preservation Alliance for Social Sciences, "Data Citations", 2012. Web page. Retrieved from: http://data-pass.org/citations.html.

Crosas, M. 2011. The Dataverse Network: An Open-source Application for Sharing, Discovering and Preserving Data. D-Lib Magazine. Volume 17.

King, G. 2007. An Introduction to the Dataverse Network as an Infrastructure for Data Sharing. Sociological Methods and Research. 36:173-199.

Novak, K., Altman, M., Broch, E., Carroll, J. M., Clemins, P. J., Fournier, D., Laevart, C., et al. (2011). *Communicating Science and Engineering Data in the Information Age. Computer Science and Telecommunications.* National Academies Press. Retrieved from: http://www.nap.edu/catalog.php?record_id=13282.

15- Microsoft Academic Search: An Overview and Future Directions

Lee Dirks[1]
Microsoft Research Connections

I would like to brief you on what we have been doing lately with the Microsoft Academic Search service. It started as a research project that has been conducted at our Beijing lab for almost eight years now. Over the course of the last eighteen months, our team in Redmond has gotten very involved in providing strategic guidance and input. Currently, we are in the process of transitioning it from a research project into an operational service that Microsoft Research will provide to the community. It will be a free academic search engine for tracking academic papers, citation links, and all the various characteristics that can be extracted from papers.

What we have been doing over the last six to nine months is working directly with open access repositories and publishers around the world to sign content agreements so we can get access to their papers. This is all about facilitating access to the papers. At present, we have 27 million papers across 14 domains, and we have another 100 million papers across more than 20 domains in the queue, pending indexing. We are going to expand our content about every three months, and are already actively evolving the site.

All of the signed content agreements that I was referencing earlier—with the various open access repositories and publishers—are to make sure that content providers are aware that we are making their data available for free. We are very interested in having the community use this service as widely as possible.

I also would like to stress that we are being as transparent as possible in talking about the number of publications and authors that we have. As soon as possible, we are going to post a list of the publishers and all the sources of this material. We are also waiting for ORCID to come online, at which point we intend to leverage their work and use their identifiers to help in the name disambiguation process.

Through the Academic Search service, people will have the ability to look at citations or publications on a cumulative or on an annual basis. The service also has some powerful visualization abilities. For example, we will have the ability to show a single author in connection with all the people that he/she has worked with in the past (e.g., co-authors).

Another thing I would like to highlight is the system's ability to drill down into fields and sub-fields. For computer science, for example, you can look at the top authors, top publications, top conferences, journals, organizations, and other characteristics. (Note that this ranking is solely based on citation counts we have calculated.) Also, you can drill down into a sub-domain of computer science and visualize, for example, publication activity using what we call the Domain Trend. We believe that Domain Trend is a very useful tool for helping researchers find co-authors, principal investigators, and even awards and people to invite to conferences. There is also the ability to do ranking across institutions and across countries.

---

[1] Presentation slides are available at http://sites.nationalacademies.org/PGA/brdi/PGA_064019.

Again, all of that information is free. We have been getting some good coverage lately, especially about some of the new functionalities of the system. Here is a recent quote from Nature[2]:

> "…*Meanwhile, Microsoft Academic Search (MAS), which launched in 2009 and has a tool similar to Google Scholar, has over the past few months added a suite of nifty new tools based on its citation metrics (go.nature.com/u1ouut). These include visualizations of citation networks (see 'Mapping the structure of science'); publication trends; and rankings of the leading researchers in a field.*"

I would like to stress the fact that the work that we are doing here is *for researchers and by researchers*. That is something that we will always keep in mind when we grow and make this a more sustainable service. We are also very interested in changing our interface and not just doing citation analysis of papers, but eventually also of data. We are very interested in conducting research projects with the community. From our perspective, Microsoft Academic Search is an open platform and we are going to be as transparent as we can about our work. We want to make sure that this service will accurately represent how science and academia work. We are going to make our domain coverage more extensive. We are also working on more partnerships. For example, we are an associate member of DataCite and we are a founding sponsor of ORCID. Finally, we are tracking these and other activities to see when and how we can integrate them into our service.

---

[2] Butler, D. 4 August 2011 *Computing giants launch free science metrics. Nature* 476, 18 (2011) (doi:10.1038/476018a).

16- Data Center-Library Cooperation in Data Publication in Ocean Science

Roy Lowry[1]
British Oceanographic Data Center

Let me start by providing some background information about the key players in the partnership that has come together to foster data publication in the ocean sciences. The Scientific Committee on Oceanic Research (SCOR) is an international non-governmental organization formed by the International Council of Scientific Unions (ICSU, now the International Council for Science) in 1957. The Committee has scientists from 36 countries participating in different working groups and steering committees. It promotes international cooperation through planning and conducting oceanographic research, and solving methodological and conceptual problems that hinder research.

The second partner organization is the International Oceanographic Data and Information Exchange (IODE). This is a data and information exchange program of UNESCO's Intergovernmental Oceanographic Commission (IOC), commenced in 1961. The main goal of this program is to establish national oceanographic data centres or coordinators in IOC member states in order to acquire, enhance, and exchange oceanographic data and information. It also aims at extending the national oceanographic data center network through training and capacity building.

The last player in this partnership is the Marine Biological Laboratory Woods Hole Oceanographic Institution (MBLWHOI) Library. The Woods Hole scientific community library has a strong interest in data publication in digital libraries. The Digital Library Archive (DLA) contains:

- WHOI archives;
- Historical photographs and oceanographic instruments;
- Scientific data, e.g., echo sounding records from WHOI research vessel expeditions;
- Technical report collections; and
- Maps, nautical charts, geologic and bathymetric maps, and cruise tracks.

The group had a series of meetings between June 2008 and April 2010 and there is another meeting scheduled for November 2011. The group's objectives are to:

---

[1] Presentation given by Sarah Callaghan and slides are available at http://sites.nationalacademies.org/PGA/brdi/PGA_064019.

- Engage the IODE data center and marine library communities in data publication issues.
- Provide a network of hosts for cited data.
- Motivate scientists through reward for depositing data in data centers.
- Promote scientific clarity and re-use of data.

However, engaging IODE data centers effectively in data publication and distribution encounters a problem of different approaches. One model is as follows.

Data can change significantly as additional value is added by the data center through metadata generation, quality control (e.g., flagging outliers), and the like.

The "best available" data are served by the data center to other users during data evolution, which means that the dataset is continually changing with no snapshots preserved or formal versioning during work-up. This makes it difficult to go back and get the same data that you got a year or six months ago.

The second model is the Digital Library Paradigm.

A dataset is a "bucket of bytes," which is:

- Fixed (checksum should be a metadata item)
- Changes generate a new version of the dataset
- Previous versions must persist
- Accessible online via a permanent identifier
- Usable on a decadal timescale (using standards such as the Open Archive Information Standard)
- Citable in the scientific literature to provide links to marine libraries
- Discoverable

To summarize these data distribution paradigm issues, the problem is to find ways for IODE data centers to engage in digital library practices while leaving current infrastructure largely intact. Change should happen gradually through evolution and not revolution. Probably the best way to do that is through pilot projects at the British Oceanographic Data Center (BODC) and WHOI.

To that end, the BODC has started a pilot project activity with a decision to establish a repository at IODE called Published Ocean Data (POD), where data will be accessible to many data centers, with technical quality control and good long-term stewardship credentials in place. The process to achieve this goal has taken longer than anticipated due to extended discussions and resource availability. However, specifications are being produced and accepted now, and the actual building of the systems will start in the fall of 2011.

As for the WHOI pilot project, the MBLWHOI library has loaded a number of datasets from the National Science Foundation's (NSF) Biological and Chemical Oceanography Data Management Office (BCO-DMO). The datasets have been associated with published journal articles. For example *dx.**doi**.org/* 10.1575/1912/4199, resolves to: https://darchive.mblwhoilibrary.org/handle/1912/4199).

The group is also working with a scientist who is submitting a paper to the American Geophysical Union in September, with a complete publishing process use case including DOI assignments to datasets supporting specific figures. These dataset citations will be incorporated in the final version of the paper, subject to publisher approval. Furthermore, talks are underway concerning incorporation of the Woods Hole Open Access Server (WHOAS) repository in an NSF proposal data management plan. Finally, this partnership also has plans for collaboration with BCO-DMO to develop an automated publication system for all data center accessions.

Let me conclude with a summary of our future plans. We will:

- Complete the pilot projects identified earlier.
- Engage other data centers in data publication through reporting our experiences and disseminating knowledge through appropriate routes, such as workshops, conferences and other publications.
- Engage SeaDataNet II when it starts later in 2011.
- Continue outreach activities to scientific, data management, and marine library communities.
- A further meeting is planned to be held in Liverpool, UK, on November 3-4, 2011.
- Expand BODC activities into an operational service.
- Develop the MBLWHOI Library BCO-DMO ingest system.

17- Data Citation Mechanism and Service for Scientific Data: Defining a Framework for Biodiversity Data Publishers

Vishwas Chavan[1]
Global Biodiversity Information Facility

I am going to focus on how we are working to resolve the issue of data citation for biodiversity data at the Global Biodiversity Information Facility (GBIF), located in Copenhagen, Denmark. For those who have not heard about GBIF, it is a multilateral intergovernmental initiative established in 2001 with 52 countries as members and 47 international organizations. GBIF's main objective is to facilitate free and open access to biodiversity data. Our data is available through a portal and currently, includes 312,000,000 data records about existence of lands and animals across the globe from over 1800 data resources that has been contributed by 342 data publishers.

Why do we think that data citation is important? We believe that data citation will encourage our data publishers to publish more and more datasets. Therefore, it will improve data discovery. It will also provide some kind of encouragement for data preservation. Furthermore, it will provide incentives to those who use the data through improving the credibility of the interpretations that are based on the data.

What is the current practice of data citation in the GBIF network? Let me explain this with an example. A user comes to GBIF's portal and searches for the term "Panthera Tigris." She gets 696 records from 37 different datasets, which are published by 31 different publishers. The current citation style just says "access through GBIF data portal" and lists out all the access points of those 37 datasets. The problem with this practice is that it doesn't tell me what was the search string unless and until I can make an explicit statement about it, how many records were retrieved, how many data publishers contributed to the retrieved data, when search was carried out, who are the original contributors of the data, and who plays what role in the process from collection to publishing of the data?

So, certainly there is a need to work around these challenges. What is needed is a data citation mechanism with a defined citation style that can provide recognition to all stakeholders involved with their roles, such as who is the producer of the data, who is the publisher, who is the aggregator, and who provided curation service to the data. Given the complexity of our network, we require cascading citations, which are citations within the citations. Furthermore, we need a data citation service whereby a publisher can go and register its citation and all documents of metadata. Finally, we need a discovery service, which resolves to the full-text citation and links to the underlying data.

One of the first things that we think we need is a best practices guide for how to cite data. For that, we require two types of recommended styles. One is related to publisher supplied dataset citation and the other is related to query based citations. The publisher supplied dataset citation would obviously need to consider the types of publishers (e.g., an individual, a group of

---

[1] Presentation slides are available at http://sites.nationalacademies.org/PGA/brdi/PGA_064019.

individuals, or an institution). We also need to recognize individual's role in creating the dataset. We need also to identify when it was first released or whether it is a one-time release or frequently updated. Also, the citation should link back to the primary URI of the dataset and then the citation itself needs to have a persistent identifier (preferably DOI) so the entire citation string can be resolved. Also, we need to consider the date of the first release, the latest updates, and the number of data records that we can actually access from a particular dataset.

Table 17-1 provides a sampling of GBIF's styles for potential citation strings or styles for the publisher supplied citations.

		Complete formulation	Short formulation
Style 1		**Publisher (individual) with one-time release of dataset**	
		Publisher (YEAR), <Title of the data resource>, <total nos. of records>, published <modes of publishing>, <Primary access point>, released on<release date>, <Persistent Identifier>.	Publisher (YEAR), <Persistent Identifier>.
Style 2		**Publisher (individual) with frequent update or release of dataset**	
		Publisher (YEAR). <Title of the data resource>, <total nos. of records>, published <modes of publishing>, <Primary access point>, first released on<release date>, <current version no. or last updated/released on (date)>, <Persistent Identifier>.	Publisher (Year first published/released -). <Version no., or last updated/released on (date)>, Persistent Identifier.
Style 3		**Publisher (group of individuals) with one time release of dataset**	
		Publisher 1, ... and Publisher n (YEAR). <Title of the data resource>, <total nos. of records>, published <modes of publishing>, <Primary access point>, released on <release date>, <Persistent Identifier>.	Publisher 1 et.al. (YEAR). Persistent Identifier.
Style 4		**Publisher (group of individuals) with frequent update or release of dataset**	
		Publisher 1, ... and Publisher n <YEAR). <Title of the data resource>, <total nos. of records>, published <modes of publishing>, <Primary access point>, first released on<release date>, <current version no. or last updated/released on (date)>, <Persistent Identifier>.	Publisher 1 et.al. <YEAR (Year first published/released -)>. <Version no., or last updated/released on (date)>, Persistent Identifier.
Style 5		**Institute/consortium (multiple contributors) with one time release of dataset**	
		<Publisher as Institution / Research Group / Consortium> (YEAR), <Title of the data resource>, <total nos. of records>, <Contributed by contributor 1(role), contributor 2 (role)..... contributor n(role)>, <published (modes of publishing)>, <Primary access point>, released on<release date>, <Persistent Identifier>.	<Publisher as Institution / Research Group / Consortium> (YEAR), <Persistent Identifier>
Style 6		**Institute/consortium (multiple contributors) with frequent update or release of dataset.**	
		<Publisher as Institution / Research Group / Consortium> <YEAR (Year first published / released -)>, <Title of the data resource>, <total nos. of records>, <Contributed by contributor 1(role), contributor 2 (role)..... contributor n(role)>, <published (modes of publishing)>, <Primary access point>,<Version no., or last updated/released on (date)>, <Persistent Identifier>.	<Publisher as Institution / Research Group / Consortium> <YEAR (Year first published / released -)>, <Version no., or last updated/released on (date)>, <Persistent Identifier>

In the case of the query based citations, where we need to have citations within citations, there are two types of citations that we think are required. One is query based citations and the other publisher supplied dataset citations. Such a citation needs to have multiple types of persistent identifiers that have been assigned or used by publishers themselves.

So, going back to the example of the user who searched for the term "Panthera Tigris", a hypothetical exemplification of this search is presented in Figure 17-1. This query based citation will resolve to complete computer citation and it can also link back to the snapshot of the retrieved data, which are cited. This is how it will look like when you resolve the DOI:

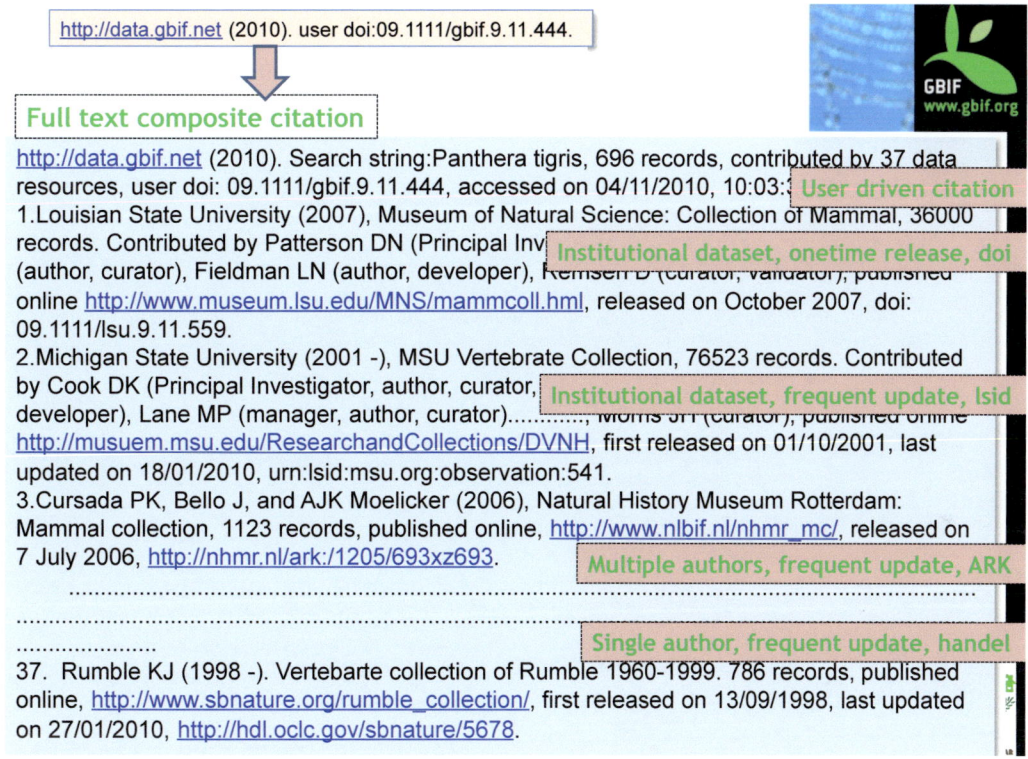

FIGURE 17-1 Hypothetical search result.

Let me conclude with a summary of the implementation and next steps. The main challenge to implement this mechanism is the complexity of data management itself. How do we make sure that all our data publishers are going to follow through the citation style that is being proposed? There is also the complexity of the data network because many publishers publish the data through more than one access point.

Therefore, we urgently need to have all these citation styles propagated in the form of a best practice guide. However, we also need to remember that there are social challenges related to updating the current practices. Finally, somebody has to come forward to run the data citation service. These are some of the challenges that we are currently trying to address.

18- How to Cite an Earth Science Dataset?

Mark Parsons[1]
University of Colorado

I represent the Federation of Earth Science Information Partners (ESIP). It is a federation of more than a hundred data centers and related organizations, predominantly in the United States. The primary sponsors of ESIP are federal science agencies NASA, NOAA, EPA, and there are several other sponsors such as NSF and USGS that are getting increasingly more involved in our work. I am going to focus my talk on best practices and guidelines of how to cite science data. I also want to mention that some of my presentation will be related to the International Polar Year (IPY), a very large international and interdisciplinary project that started to work on these issues.

There is a lot of input going into the process of creating data citation and attribution guidelines at ESIP. We hope that these guidelines will be adopted by the general assembly in January of 2012[2]. The main purposes of data citation as we see them are:

- Credit for data authors and stewards.
- Accountability for creators and stewards.
- Track impacts of the dataset.
- Assist data authors in verifying how their data are being used.
- Aid reproducibility of research results through a direct, unambiguous connection to the precise data used.

The last purpose is the primary, most important purpose and the most difficult to achieve. I also want to note that we see citation as a reference and a location mechanism, but not as a discovery mechanism, per se.

Data citation in the earth sciences is currently done using one of these approaches or styles:

- Citation of traditional publication that actually contains the data, e.g., a parameterization value.
- Not mentioned, just used, e.g., in tables or figures.
- Reference to name or source of data in text.
- URL in text (with variable degrees of specificity).
- Citation of related paper (e.g., the UK Climate Research Unit recommends citing their well-known surface temperature records using two old journal articles which do not contain the actual data or full description of methods)
- Citation of actual data set typically using recommended citation given by data center.
- Citation of data set including a persistent identifier or locator, typically a DOI.

---

[1] Presentation slides are available at http://sites.nationalacademies.org/PGA/brdi/PGA_064019.
[2] The Guidelines were adopted in January.

The National Snow and Ice Data Center (NSIDC) distributes a variety of different snow cover products derived from the Moderate Resolution Imaging Spectrometer (MODIS). The results of a quick analysis of how many scientific papers mention use of "MODIS Snow Cover Data" (according to Google Scholar) and how often the data sets themselves are formally cited shows a huge disparity, illustrating the infrequency of proper data citation in practice. Moreover, the lack of data citation standards introduces the possibility that informal references to data do not point to the exact data set actually used.

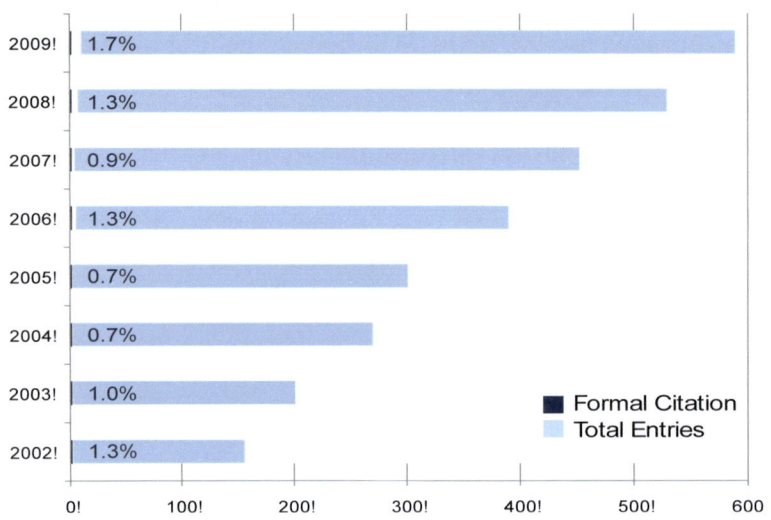

FIGURE 18-1 MODIS snow cover data in Google Scholar.

There are a number of data citation guidelines available to scientists. These include the ones from the International Polar Year and DataCite project. Also, institutions such as NASA and NOAA request acknowledgments. Overall, approaches range from specific data citation, to general acknowledgment, to recommending citing a journal article or even a presentation. This is reflected also in the results of this study titled "Data Citation in the Wild" by Enriquez et al. (2010):

> We found that few policies recommend robust data citation practices: in our preliminary evaluation, only one-third of repositories (n=26), 6% of journals (n=307), and 1 of 53 funders suggested a best practice for data citation. We manually reviewed 500 papers published between 2000 and 2010 across six journals; of the 198 papers that reused datasets, only 14% reported a unique dataset identifier in their dataset attribution, and a partially-overlapping 12% mentioned the author name and repository name. Few citations to datasets themselves were made in the article references section.[3]

This shows clearly that the data author is not being fairly credited.

---

[3] Available at: http://openwetware.org/wiki/DataONE:Notebook/Summer_2010.

In terms of measuring the impact of a data set, there are some measurement issues that make this process a bit challenging[4]. For example, data are not used in isolation. Often different data are combined or used with models and other analytic techniques. Also, impacts may be indirect (i.e., resulting from development of information, papers, tools, etc. that relied on derived data or products); they may be delayed (i.e., months or years for a peer-reviewed publication to be released, or a decision to be made and implemented); they may be unexpected (e.g., a new scientific discovery or a novel application of data collected for a different purpose); or they may be hard to compare (e.g., in scientific, economic, or ethical terms). Nevertheless, it is still important to try to track the use and impact of a data set because we need to justify investment in data acquisition, maintenance, distribution, and long-term stewardship. We also need to help the community become more effective and efficient in data management and use.

There are different possible citation metrics. These include:

Qualitative

- Examples of data use and impacts in key papers, discoveries, decisions.
- Assessment of broader impacts such as influence of data on attitudes and thinking (e.g., the Apollo 8 image of the Earth).

Quantitative

- Counts of papers that cite data in peer-reviewed journals.
- Weighted indicators of data citations (e.g., type and quality of citation, impact of journal).

Quantitative and Qualitative:

- Number of data citations in top peer-reviewed scientific journals and key reports by decision-makers.
- Data usage in other peer-reviewed journals, textbooks, reports, magazines, documentary films, online tools, maps, blogs, twitters, and the like.

However, as Heather Piwowar notes, tracking dataset citations using common citation tracking tools does not work. Traditional fields, such as author and date, are too imprecise and the Web of Science, Scopus, and other scientific publisher tools do not handle identifiers.[5]

I think that we need two basic strategies. One is that archives and data centers need to provide consistent and precise recommendations on how their data should be cited. The other strategy is more of the social strategy trying to get the publishers and the educators on board with the whole concept of data citation. I am going to focus on the first strategy in this presentation.

---

[4] See also Chen, R. S. and Downs, R. R. (2010). Evaluating the Use and Impacts of Scientific Data. National Federation of Advanced Information Services (NFAIS) Workshop, Assessing the Usage and Value of Scholarly and Scientific Output: An Overview of Traditional and Emerging Approaches. Philadelphia, PA, November 10, 2010. http://info.nfais.org/info/ChenDownsNov10.pdf.
[5] See Piwowar's blog at http://researchremix.wordpress.com/2010/11/09/tracking-dataset-citations-using-common-citation-tracking-tools-doesnt-work/.

Below is the basic ESIP data citation model shown in contrast to the DataCite guidelines available at the time.

Per DataCite:
Creator. Publication Year. Title. [Version]. Publisher. [Resource Type]. Identifier.

Per ESIP:
Author(s). Release Date. Title [version]. [editor(s)]. Archive and/or Distributor. Locator. [date/time accessed]. [subset used].

I will use the rest of the talk to describe some of these differences and why we think they are important.

The first difference is that ESIP explicitly allows the recognition of roles other than the data creator or author. We call this "editor", but there are multiple data management roles that might be captured. Whether or not they are appropriate can be open to question, but this approach gets a lot of traction with data stewards because particularly in earth sciences, data stewards frequently may have a significant role in developing and compiling the data sets and sometimes doing some quality control. They have similar levels of credit and accountability as the original authors do and I think that is important to recognize. For example, in the example below, the data authors were the designers of a large field experiment. The editors were responsible for managing the process of entering field data from notebooks, conducting manual and automated quality control, determining data formats, writing documentation, and so on.

Cline, D., R. Armstrong, R. Davis, K. Elder, and G. Liston. 2002, Updated 2003. CLPX-Ground: ISA snow depth transects and related measurements ver. 2.0. Edited by M. Parsons and M. J. Brodzik. Boulder, CO: National Snow and Ice Data Center. Data set accessed 2008-05-14 at http://nsidc.org/data/nsidc-0175.html.

Another concept I want to present is the notion of the identifier versus the locator. The easiest way for us to understand these concepts is probably to look at the human example.

Human ID: Mark Alan Parsons (son of Robert A. and Ann M., etc.).

- Every term defined independently (only unique in context/provenance).
- Alternative like a social security number requires a very well controlled central authority.

Human Locator: 1540 30th St., Room 201, Boulder CO 80303.

- Every term has a naming authority.

Data Set IDs: data set title, filename, database key, object id code (e.g., UUID), etc.

Data set Locators: *URL, directory structure, catalogue number, registered locator (e.g. DOI), etc.*

If we look at the human ID, every term is defined independently and it is only unique in a certain context. We could use a title in combination with a location to find the relevant person, but it would not necessarily be the right person. S/he might have retired. We could use his or her identifier but that may not describe her location or the person may have moved. This may be simplistic, but I see this same situation with data. There are data set IDs, some of them are informal, like dataset titles, and others are very formal, like a UUID. There are also data set locators like URLs or some registry based system like DOIs.

The point is that the locator and identifiers are different things, but sometimes locator can be used as an identifier (e.g., the person working in this position at this address). Hence the general use of the term "identifier" such as in DOI, is better described as a locator.

Indeed it is the registration of the location information in the DOI scheme that makes it attractive to groups like DataCite:

> One of the main purposes of assigning DOI names (or any persistent identifier) is to separate the location information from any other metadata about a resource. Changeable location information is not considered part of the resource description. Once a resource has been registered with a persistent identifier, the only location information relevant for this resource from now on is that identifier, e.g., http:// dx.doi.org/10.xx. [6]

Duerr et al (2011)[7] conducted an assessment of identification schemes for digital earth science data as summarized in this diagram I adapted from their paper.

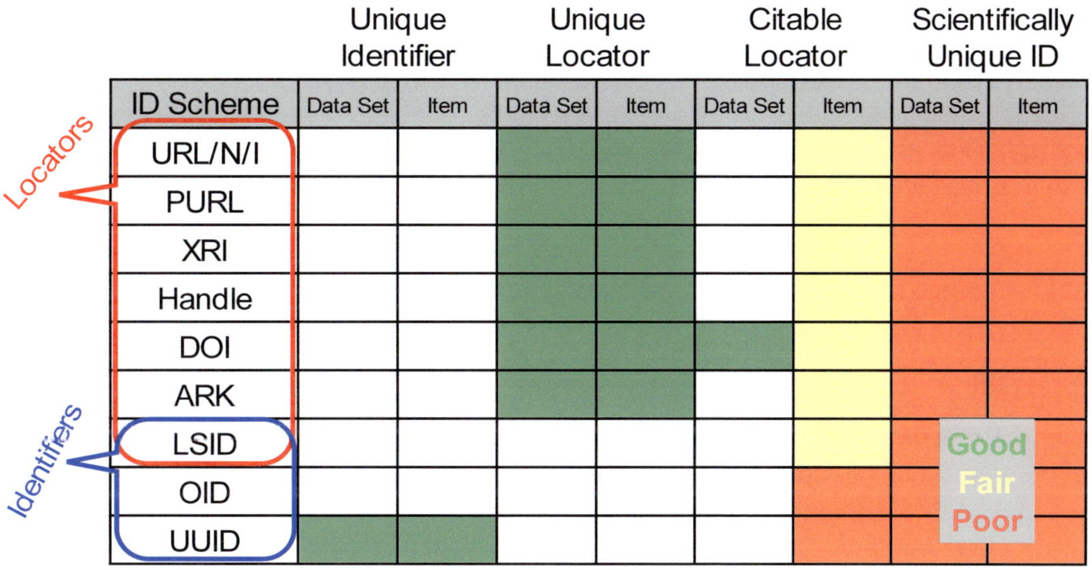

FIGURE 18-2 Assessment of identification schemes for digital earth science data.
SOURCE: Duerr et al (2011).

---

[6] DataCite Metadata Scheme for the Publication and Citation of Research Data, Version 2.2, July 2011.
[7] Duerr, R., R. Downs, C. Tilmes, B. Barkstrom, W. Lenhardt, J. Glassy, L. Bermudez, and P. Slaughter. 2011. On the utility of identification schemes for digital earth science data: an assessment and recommendations. *Earth Science Informatics*: 1-22. http://dx.doi.org/10.1007/s12145-011-0083-6.

The figure summarizes how different identifiers are more suitable for different purposes, and that often it depends on whether the scheme is actually a locator or an identifier. (Note that the LSID is a locator; but also the ObjectID part of it is an identifier and most people use a UUID for the ObjectID part of it.) Also the ARK could be considered a bit better than the rest of the locators because it has additional trust value, but the DOI stands out as the most appropriate locator for citation.

Why the DOI? Although the DOI is not perfect, it is well understood and accepted by publishers, and DataCite is working with Thomson Reuters to get data citations in their index. This broad acceptance gives DOIs a small edge, but, there are still some issues that need to be resolved. For example, what is the citable unit that should be assigned a DOI? Is it a file or a collection of files and, if so, how many? How do we handle different versions? When does a new version get a new DOI? How do we handle data that have been retired and deleted? Does their DOI persist? What does it point to?

Overall, we believe these issues can be largely resolved by, a well-defined versioning scheme, good tracking and documentation of the versions, and due diligence in archive and release practices. So, it is not a technical problem so much as a social problem demanding good professional practices.

Here are some initial suggestions on versioning and locators. At my data center, we did a study looking at different types of data, from satellite data, modeling output, historical photographs, to interviews and transcripts. We have the notion of a major version, a minor version, and an archive version. The archive version is not publicly available, it is just for us to track any changes in the archive. What constitutes a major or a minor version has to be done on a case-by-case basis. An individual steward has to work with their providers to figure it out, but in general, something that affects the entire dataset is going to be a major version. A small change such as changing a land mask might be a minor version.

DOIs should be assigned to major versions. Old DOIs for old versions should be maintained even if the data are no longer available. The old DOIs should point to some appropriate page that explains what happened to the old data if they were not archived. The older metadata record should remain with a pointer to the new version and with explanation of the status of the older version data. Major and minor versions (after the first version) should be exposed with the data set title and recommended citation. And while minor versions don't get a new DOI, they should be explained in documentation, ideally in file-level metadata. Finally, applying UUIDs to individual files upon ingest aids in tracking minor versions and historical citations.

The last difference between ESIP and DataCite is the inclusion of "subset used" This is the concept of micro citation, which may be the most challenging aspect of data citation. In conventional literary citation this might take the form of citing a passage in a book and referencing a page number. We all know how to deal with page numbers in a book. But, how do we do it in a data set? Maybe we can put an identifier to it. If we have a particular query, we could capture the query and maintain sort of a query ID. Those kinds of technical approaches are probably the way forward but that is not the way the vast majority of group science data is managed today. So instead, we consider the concept of a structural index. This is similar to citing "chapter and verse" in a sacred text.

The key question then is what structure or structures can we use to organize data collections that might be common across earth sciences? The basic assumption of a "chapter-verse" style of reference is that there is a canonical version of data set. This is also assumed in the approach using the Unique Numerical Fingerprint. Unfortunately, most earth science data lack a canonical version. For example, data could be in different digital formats, where the contents are scientifically equivalent, but they are not identical because of the different formats. Therefore, we need to refer to "equivalence classes" not canonical versions, although we cannot deny the human readability of the chapter-verse style approach.

We probably need both approaches. We need the "chapter and verse" that makes sense to people and is easily conceived and communicated between people, but then we still need the precise location and identity of that rather mutable verse represented in a way that computers can readily understand and be precise about, i.e., the identifier. And then we cannot forget the fact that we have billions if not trillions of "verses" or "granules" that we are dealing with. Our human approach needs to make sense at a high level of aggregation, while the computer approach needs to handle the volumes and precision.

In earth science data, space and time can often serve as a structural index. We can simply refer to a spatial and temporal subset of the data. We might also consider what the Open Archives Information System Reference Model[8] calls archive information units: An Archival Information Package whose Content Information is not further broken down into other Content Information components, each of which has its own complete Preservation Description Information.

Neither of these approaches is fully satisfactory, but following are some examples of doing it as best we can:

Hall, Dorothy K., George A. Riggs, and Vincent V. Salomonson. 2007, updated daily. MODIS/Aqua Snow Cover Daily L3 Global 500m Grid V005.3, **Oct. 2007- Sep. 2008, 84°N, 75°W; 44°N, 10°W.** Boulder, Colorado USA: National Snow and Ice Data Center. Data set accessed 2008-11-01 at doi:10.1234/xxx.

Hall, Dorothy K., George A. Riggs, and Vincent V. Salomonson. 2007, updated daily. MODIS/Aqua Snow Cover Daily L3 Global 500m Grid V005.3, **Oct. 2007- Sep. 2008, Tiles (15,2;16,0;16,1;16,2;17,0;17,1).** Boulder, Colorado USA: National Snow and Ice Data Center. Data set accessed 2008-11-01 at doi:10.1234/xxx.

Cline, D., R. Armstrong, R. Davis, K. Elder, and G. Liston. 2002, Updated 2003. CLPX-Ground: ISA snow depth transects and related measurements, Version 2.0, **shapefiles**. Edited by M. Parsons and M. J. Brodzik. Boulder, CO: National Snow and Ice Data Center. Data set accessed 2008-05-14 at doi:10.1234/xxx.

We have not solved all the issues related to data citation and attributions, but we believe that approximately 80 percent of citation scenarios for 80 percent of Earth system science data can be addressed with basic citations, i.e., *[(Author(s). ReleaseYear. Title, Version. [editor (s)].*

---

[8] CCSDS (Consultative Committee for Space Data Systems). 2002. *Reference Model for an Open Archival Information System (OAIS) CCSDS 650.0-B-1 Issue 1.* Washington, DC: CCSDS Secretariat.

Archive. Locator. [date/time accessed]. [subset used]], and reasonable due diligence. We need to move forward with this now and not wait for the perfect solution.

Finally, as we go forward, I think that the concept of scientific equivalent is ripe for study and that we are beginning to look at the notion of how content equivalence and provenance equivalence can serve as a proxy for scientific equivalence. That is a big research question, but it should not stop us for moving forward on the citation issue in general. I want to emphasize that we can do something about data citation now and we should.

19- Citable Publications of Scientific Data

John Helly[1]
University of California at San Diego

This presentation focuses on what we have learned from a history of developments in scientific data publication that began in 1993 and continue today. The first data publication work at the San Diego Supercomputer Center started in 1993 related to natural resource management in San Diego Bay and evolved to an activity with the Ecologic Society of America (ESA) in order to solve some problems related to the preservation of long-term ecological data. These data were at risk of being lost, but in 1998 we set up a website that was designed for publishing data papers by the ESA. This effort then led to a letter in *Nature*[2], which suggested that the scientific community should raise data collections to the status of citable entities in journals. This was followed by an ACM publication in 2002[3] and several other publications related to scientific data publication in the earth sciences and scalable models of data sharing. This meant that we were able to distill some basic principles and requirements for systems. These are the design principles that we employ in systems now in operation as well as new systems across disciplines and domains.

The three earliest digital library systems in continuous operation since their inception are:

1- The Scripps Institution of Oceanography (SIO) SIOExplorer, since 2001.
2- The Site Survey Databank (SSDB) for Integrated Ocean Drilling Project (IODP), since 2003.
3- The National Science Foundation Center for Multi-scale Modeling of Atmospheric Processes (CMMAP) Digital Library project in the atmospheric science, since 2005.

These systems are designed to deal with data up to the multi-petabyte range for data storage and transportation requirements. From these developments we have learned how to change the workflow for scholarly publication to achieve the goal of citable scientific data. The basic workflow for scientific scholarly research starts with collecting data, doing the research, writing and publishing a manuscript for which some of the people get credit for it through citations. Within the past few years, it has become possible for individuals to obtain the authority to issue digital object identifiers (DOIs). Previously this was an authority available only to commercial publishers. This new capability allowed us to introduce the use of DOIs for data to this workflow and make citable data publication a reality.

---

[1] Presentation slides are available at http://sites.nationalacademies.org/PGA/brdi/PGA_064019.
[2] J. Helly. New concepts of publication. Nature, 393, 1998.
[3] J. Helly, T. T. Elvins, D. Sutton, D. Martinez, S. Miller, S. Pickett, and A. M. Ellison. Controlled publication of digital scientific data. CACM (accepted October 3. 2000), May, 2002.

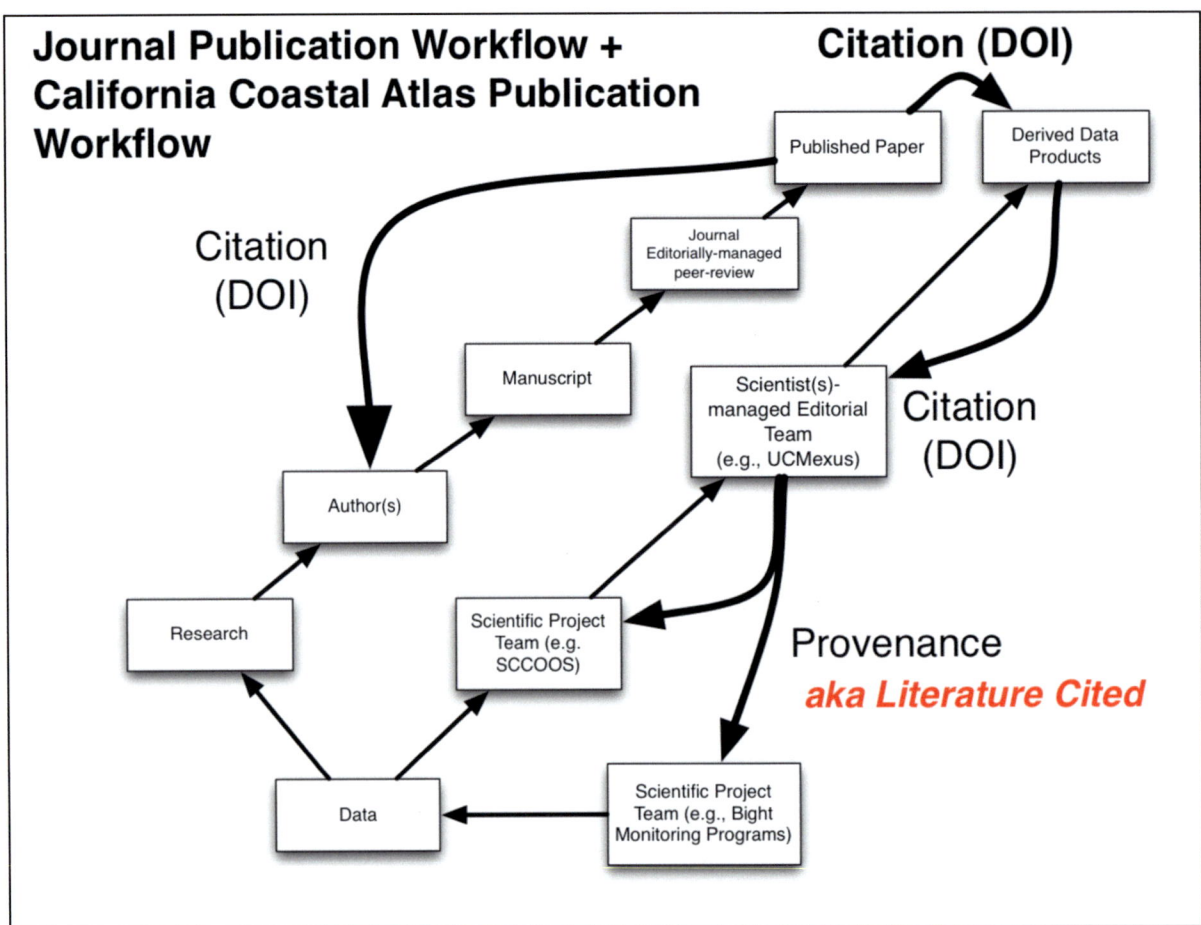

FIGURE 19-1 Basic scholarly workflow paralleling the new corresponding workflow for data publication. SCCOOS and UCMexus are acronyms pertaining to specific projects.

We kept the same basic workflow with a path for data paralleling the manuscript path. The key here is to develop the training necessary to teach these steps to graduate students and expert scientists to ensure progress in this area for a number of reasons. Only scientific experts can ensure data quality and provide sufficient metadata to enable this process. The federal agency archival requirements for data developed under federal grants are clearer than previously, but we need some incentives as well. There are also financial issues to deal with: How are long-term archives to be supported?

Non-interoperability of DOIs from different systems is also a looming problem. Recent information has come to light that the main DOI providers for data are not interoperating. This is a problem because the whole concept of changing the workflow hinges on the ability to resolve the DOI issues across the different domains and publishing systems. The scientific community may need another solution that fully realizes the value of DOIs and warrants the effort to use them. It looks like many of the old players in the publication industry are moving to "wall-off" what they perceive as their intellectual property by sequestering their DOI cross-referencing.

Let me now talk about the California Coastal Atlas. It is designed for data publication, with a focus on developing methods and training people to do high-quality scientific data production, primarily in the geospatial data area. The model is scalable by design. We know that science proceeds through research projects and that these projects have finite life times. The key people are the Principal Investigators, the research managers sponsoring the projects, and the other people who are doing the work. So, through cooperation between the chief editor of the Atlas and the different projects teams, the projects agree to do their data management according to the Atlas conventions and standards. By modifying that workflow slightly, though not dramatically, we were able to provide a platform for those scientific projects to have high-quality data end products.

The current projects are:

- UCMexu: Declining Oxygenation and pH of the Eastern Pacific Margin.
- US Navy: A Methodology for Assessing the Impact of Sea Level Rise on Military Installations in the Southwestern United States.
- California Environmental Data Exchange Network: the 303D-listing Dataset.
- The California Spatial Data Infrastructure.

We believe that the real keys to success in this process are a set of factors that can be summarized as follows:

- Changing scientific workflows in familiar, but powerful ways to attribute high-quality data to the authors.
- Incentivize researchers to modify their existing workflows only slightly and provide tools to do it.
- Integration into a well-established and trusted system of scholarly publication.
- Providing the basis for protecting intellectual property rights.

The figure below provides the visual representation of our approach to automate the production of metadata. The intermediate products that are generated automatically include a bibliographic reference file, a metadata interchange file that talks only to OAI- PMH, and then the basic underlying metadata or the data content in the form of what we call an arbitrary digital object.

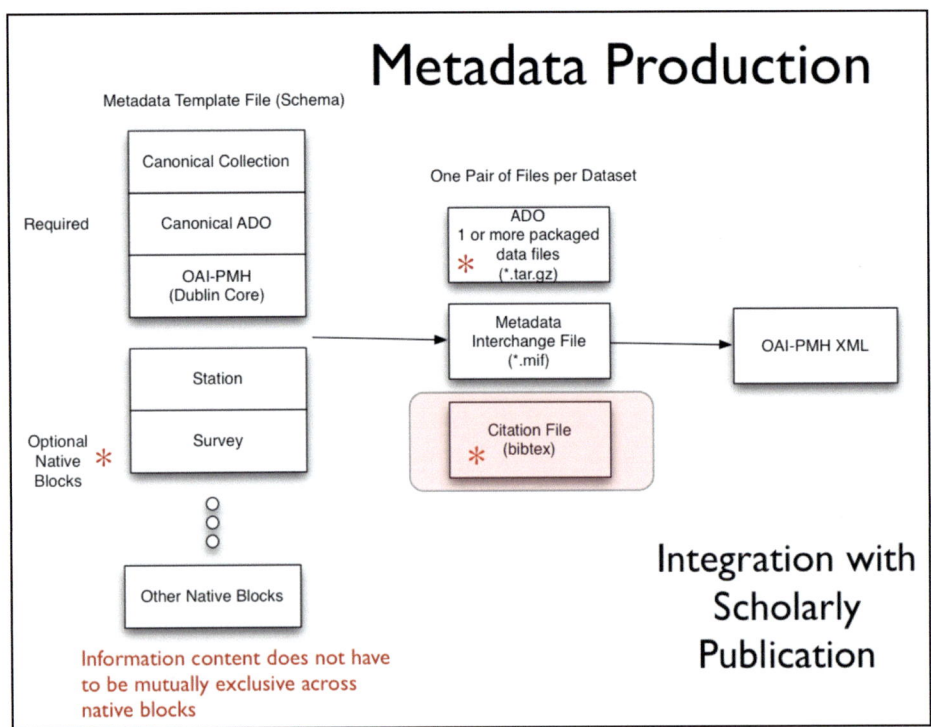

FIGURE 19-2 Metadata production process emphasizing the modular nature of metadata organization to support the minimal needs for cataloging as well as the disciplinary needs for re-use of the data.

We use conventional tools (LaTex/BibTex) that have seen a resurgence in the past five years to produce the content of the Atlas and to ingest the bibliographic reference information using tools like BibTex, so that the data underlying an image, for example, could be directly cited within the context of the document in the California Coastal Atlas.

The editorial policy is probably the most confusing part, especially in terms of how it is actually done. The following figure attempts to depict it.

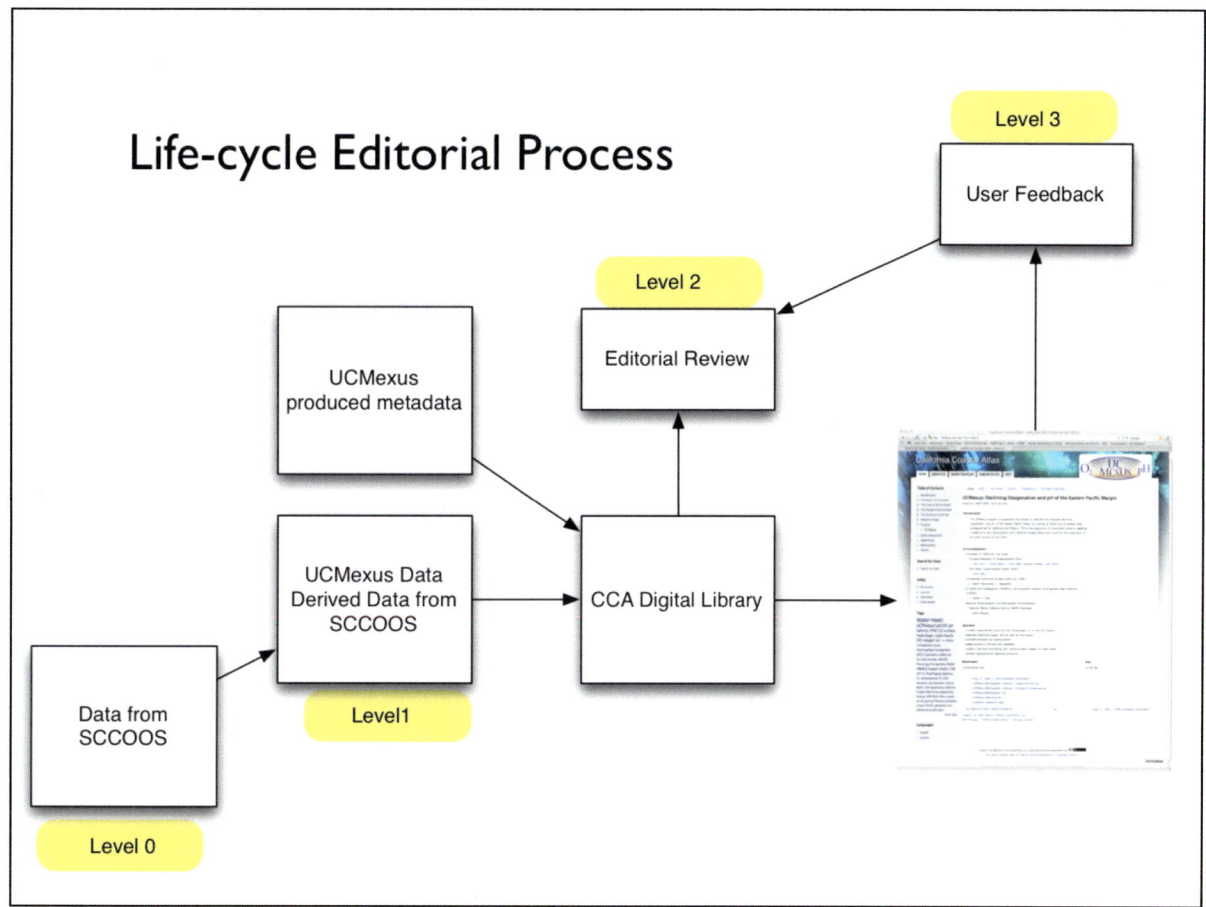

FIGURE 19-3 The editorial workflow organized into levels with requirements to transition from one level to another. Level 0 is raw data. Level1 is data that has been quality controlled and provisioned with metadata. Level 2 data is data that has been through peer-review and Level 3 data has been used by others and may be combined with other data.

We define levels of data in the form of state machine transitions, since there are requirements for going from level zero to level one and then to level two. There is always a question of managing derived data, how to combine and track it and that is where DOIs play a powerful role. There is an on-going question of user feedback when data anomalies are found in subsequent use and the project that produced the data has ended. How do anomalous reports get factored back into the maintenance of the data collection?

I will conclude with this set of editorial requirements for data publication, which are essentially the instructions to authors. With editorial guidance, data authors should provide:

- Derived data products in CCA-conforming data format and packaging;
- CCA-conforming metadata (fully-provenanced);
- Procedural software for reading the data object;

- Corresponding output listing for verification of data contents;
- Metadata for obtaining a Digital Object Identifier;
- Manifest with summary description (e.g., README) describing what is contained in the arbitrary digital object; and
- Licensing statement.

20- The SageCite Project

Monica Duke[1]
The University of Bath

I am the project manager at "SageCite," a project funded by JISC (the expert organization on information and digital information for education and research in the United Kingdom) through the Managing Research Data program in the United Kingdom. The project focuses on disease network modeling within Sage Bionetworks.

SageCite was funded between August 2010 and July 2011 to develop and test a citation framework linking data, methods, and publications. The domain of bio-informatics provided a case study, and the project builds on existing infrastructure and tools: myExperiment and the Sage Commons. Sage Commons is an initiative of Sage Bionetworks to build a platform to share data in bio-informatics. Citations of complex network models of disease and associated data will be embedded in leading publications, exploring issues concerning the citation of data including the compound nature of datasets, description standards, and identifiers. The project has international links with the Concept Web Alliance and Bio2RDF. The partners are UKOLN, the University of Manchester and the British Library (representing DataCite), with contributions from *Nature Genetics* and *PLoS*.[2]

The project was structured through a number of work packages comprising:

- Review and evaluation of options and approaches for data citation.
- Understanding the requirements for citing large-scale network models of disease and compound research objects.
- Demonstration of a citation-enabled workflow using a linked data approach. http://blogs.ukoln.ac.uk/sagecite/demo/
- Benefits mapping using the KRDS2 (Keeping Research Data Safe) taxonomy. http://www.beagrie.com/SageCite-KRDS_BenefitsWorksheet.pdf
- Technical and policy implications of citation by leading publishers. http://blogs.ukoln.ac.uk/sagecite/publisher-interviews/
- Dissemination across communities (bio-informatics and research and information communities).

Sage Bionetworks is a non-profit organization located in Seattle, WA that is creating resources for community-based, data-intensive biological discovery. Their work is motivated by the belief that it is necessary to have community-based analysis to build accurate models. They are also driven by the fact that no one single body has all the data required to build accurate models, so different stakeholders come together and contribute. Sage Bionetworks provides the data infrastructure, the culture, and the norms to make this happen.

---

[1] Presentation slides are available at http://sites.nationalacademies.org/PGA/brdi/PGA_064019.
[2] See project description at: http://blogs.ukoln.ac.uk/sagecite/.

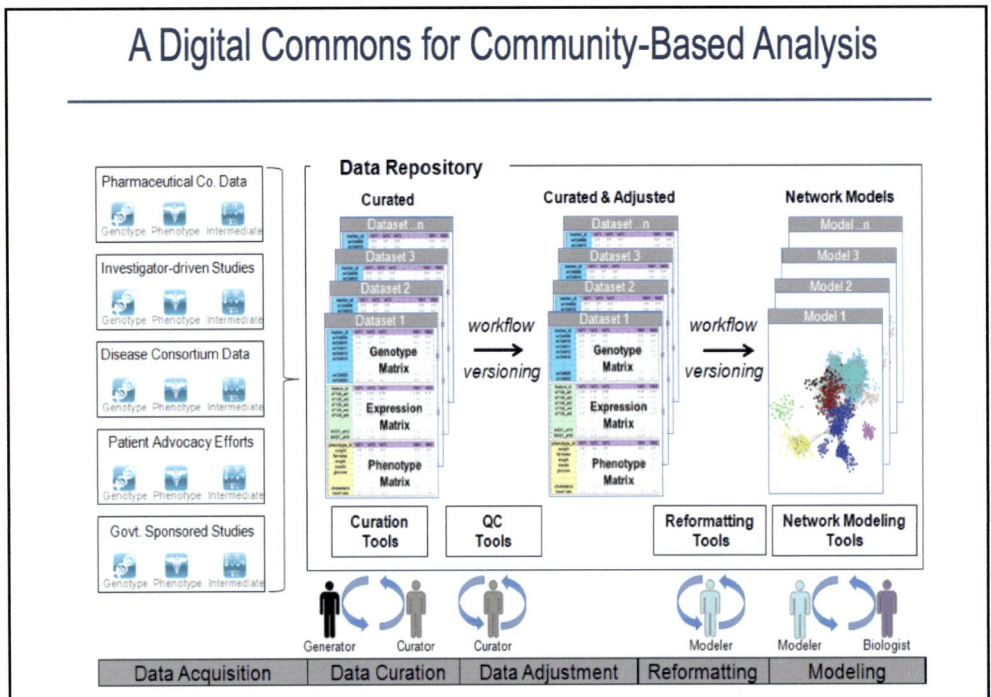

FIGURE 20-1 A digital commons for community-based analysis.

There are different types of data that the platform hosts and they come from different sources. For example, data can be obtained from pharmaceutical companies, disease consortia, investigators, patient advocacy organizations, and from government sponsored studies. There are seven stages in the data processing pipeline. The pipeline requires as input a combination of phenotypic, genetic, and expression data that need to be processed to determine a list of genes associated with diseases. The following figure shows an (idealized) description of these steps, each of which is likely to be performed by a different scientist who specializes in that area. One scientist acts as the project lead.

FIGURE 20-2 Stages in lifecycle.

Stage 1: Data Curation—This consists of basic data validation to ensure integrity and completeness of the data (although some files use common formats, others have considerable variety.) The datasets include microarray data and clinical data. This step ensures that the format of the data is understood and the required metadata is present.

Stage 2: Statistical QC— Actual values in data are validated for quality to check for experimental artifacts. The checks made are dependent on the type of data set and involves the use of R scripts (for statistical computing) or specialized gene analysis tools (like "Plink."). The output is a normalized dataset.

Stage 3: Genomic Analysis— This involves identifying regions in the genome associated with clinical phenotypes and other molecular traits. The Sage Genetic Analysis Pipeline, which consists of a set of R and C programs, is used. Statistical analysis is applied to identify interesting loci significantly associated with specific phenotypes (e.g., clinical phenotypes such as cQTL).

Stage 4: Network Construction— This stage focuses on building a network using a statistical technique to capture how biological entities, such as genes, are related to each other. Networks can contain up to 100 thousand nodes. In the network, nodes represent biological entities of some type (a gene, a protein, or even a physiological trait) and edges represent relationships between pairs of nodes. The output could be a correlation network (undirected graph) or a Bayesian network (directed, acyclic graph).

Stage 5: Network Analysis— This involves examining the network to determine how its function can be modulated by a specific subset of biological nodes. The output may be a list of genes or a sub-network. The networks from the previous steps are analyzed using techniques like Key Driver Analysis to determine a subset of interest.

Stage 6: Data Mining— A report validating claims from network analysis is produced by a domain specialist with knowledge of the study domain. This stage uses resources from the literature and public databases to assess the predictions. The information is used to annotate network models to build the case for the involvement of genes in the functioning of the network.

Stage 7: Experiment Validation— In the final stage, laboratory experiments are devised and performed to test the claims of the model. Validation is not carried out at Sage Bionetworks, but is completed in partnership with Sage Bionetworks collaborators.

Such a complex process presents challenges for reproducibility and citation. Data curation is required as a first step to do basic data validation to ensure integrity and completeness, and to ensure that the format of the data is understood and the required metadata is present. Agreed standards are also required for data sharing. We have to make sure that the data from different sources can be described, shared, used, and make the discovery process easier.

The project has employed the workflow tool, Taverna[3], which helps to document the data processes and enables the workflow to be re-enacted. The workflow can also be registered with a Digital Object Identifier (DOI). Capturing the workflow and assigning an identifier supports better citation because the cited resource is more re-usable, and strengthens the reproducibility and validation of the research.

Finally, we can describe the challenges for using data citation with the purpose of giving attribution and supporting reproducibility within this specific context. The challenges for attribution include:

---

[3] http://www.vimeo.com/27287109.

- Preserving a link with the original data.

    – The data that enters the processing pipeline originates from several sources with different methods of identification of that data (or none). Some discipline-based repositories have their own identifiers that are culturally the norm within the discipline, but may not be well-known within other communities, or may not fit in technically with global identifier infrastructures.

    – Creating Bi-directional links.

    It is not sufficient to keep links which go in one direction only from processed data to the originating data. The originators of the data (e.g., discipline repositories) would also like to track usage. Therefore links need to be maintained in both directions. However, systems of notification of usage and tracking have not yet been developed.

- Attributing data creators.

    – Identifying the party that created or contributed the data may not be straightforward and may have confidentiality issues (e.g., where medical data identifies specific populations). The situation is made more complex through developments of sites like PatientsLikeMe, where individuals are choosing to contribute their data. The range of entities and individuals who expect to be credited can be expected to grow and identifying new categories of data contributors (such as the individual patient) will create new challenges.

- Defining creation of new intellectual objects, e.g., a curated dataset.

    With a complex process the community needs to agree what represents a new intellectual object that should be attributed. A curated dataset represents a significant input from the curator to make the object usable, but is the curated dataset a new distinct object that should be attributed and identified separately to the original data?

- Cultural challenge in recognizing non-standard contributions; micro-attribution.

    Traditionally there has been emphasis on publications as a measure of contribution for the purposes of career advancement and peer recognition. A culture change is required if other categories of contribution (such as curation effort and data sharing) are to be recognised. Unless these contributions are recognised there will be little motivation to put in the effort to attribute them and create citation mechanisms around them. Micro-attribution is a developing idea in data citation to recognise smaller contributions and was used in the description of genetic variation in a paper in *Nature Genetics* in March 2011.[4]

- New metrics.

---

[4] http://www.nature.com/ng/journal/v43/n4/full/ng.785.html.

As new types of contribution are recognised complementary metrics and mechanisms for measurement will be required. Communities will need to decide what should be measured and services will need to be devised to track data citation and measurements.

- Identification of contributors.

With multi-step processes where individuals with different roles contribute, methods will be needed to describe the role of individuals and their contributions, particularly if non-traditional contributions such as data curation, data processing, data analysis, software, or process development are to be attributed.

Reproducibility:

- Identification and granularity.

    – Discipline identifiers, global identifiers.

    SageCite has taken a workflow capture approach to preserving the steps of the process to make it reproducible. When assigning identifiers for citation purposes decisions are required to decide at which level of granularity unique identifiers should be issued. Is it sufficient to identify the work flow or do individual steps need to be assigned their own identifiers? When should discipline identifiers be incorporated and how are these associated with identifiers assigned from a global system?

    – How much value has been added since the data entered the workflow?

    – One argument for deciding when a new identifier is required is to assess the value added to the data since it has been in the pipeline. To ensure reproducibility and provenance tracking, links need to be kept between value-added versions which have acquired a new identity in the pipeline and the original data.

- Identifying processes and software.

For the purposes of reproducibility it is not only the data that needs to be identified and cited. The tools and the workflow applied need to be referred to and accessed. The exact details are not always recorded and although some generic tools (such as R) are sometimes cited, the specific scripts used must be curated in order to become citable objects.

DISCUSSION BY WORKSHOP PARTICIPANTS

Moderated by David Kochalko

PARTICIPANT: This session reminded me that the history of documentation is full of powerful systems that died because they were too much work to operate.

MR. KOCHALKO: I think it is difficult to draw meaningful distinctions between locators and identifiers. I also think it is important to understand that when some providers change a version number they will also change the identifier, which maintains parity so that each version of a dataset can both be located and uniquely cited. When providers refuse to issue new identifiers, they make it difficult to associate the version of work with its location and a unique citable identifier. I think that these factors have to be accommodated and that it is still possible to maintain version histories, which is a really fundamental.

MR. PARSONS: Quite honestly, most earth science data are not well versioned currently. What we have found is that an accurate citation is highly coupled with provenance and we, as a community, are just now beginning to fully address provenance. My data center recently got some money to develop a so-called climate data record, which is meant to be the gold standard of a long time series, in this case, brightness temperatures measured from passive microwave sensing satellites. What we discovered is that the dataset that can be perfectly reproduced was actually not the best dataset because scientists had made decisions over its 30-year history that they were not necessarily documented in a way that could be reproduced by a machine. My point is that the provenance is really key and it is a developing field.

PARTICIPANT: The major versions approach is good, but the other approach, which I believe the British Oceanographic Data Center (BODC) is using, is periodic snapshots. Either way, it is not an identifier.

DR. CALLAGHAN: At the BODC and by extension the rest of the UK National Environmental Research Centre data centers, when we post a DOI on a dataset, we are saying that it is frozen in time and will not change as far as we can possibly manage it. If a dataset is still being updated, it will not have a DOI. It will still be accessible and citable, using URIs and URLs, but it will not have a DOI association.

PARTICIPANT: When you say a "dataset", do you mean, for example, the time series of the history of the Earth's temperature?

DR. CALLAGHAN: For those situations where we do have an ongoing time series, we divide it into decades or years or even months, if appropriate. That kind of dataset, however, is picked because once you have recorded the data, they are not going to change. One will not go back after the fact to change what is in that particular time period, unless there is a major problem. In that case, you have to redo the dataset or revise the calibration and then you issue a new DOI with a new version.

PARTICIPANT: The data processed by DataCite there are freely accessible by anybody. What happens if the commercial data curator decides to get out of the business? What happens to the data?

DR. BRASE: First of all, we believe that the access to the data should be free of charge, but there is no strict rule that they always should be free of charge. We therefore work together with data centers that need to get some compensation for the data and we encourage them to make access as free as possible. The data centers do not seek to make profit, but there are some that do provide access to the data for a fee, or the data are only available to the members of some institution.

Now to your question about what happens to the data? That is always an issue and that is a situation for which we still have not found a perfect solution. One of the good things about assigning an identifier to a dataset is that you always can ensure that when somebody references the DOI name of a dataset that is no longer available, they will not get a 404-error, but they will receive a page describing that this dataset is no longer available and where the last known version can be accessed. This is always a possibility, but the idea of DataCite is that, ideally, if we would have a situation where one data center would cease to exist, we would try to find other data centers to take over the data and ensure that DOI name refers to the current version of the data. If that does not work, then we would direct the DOI to a page describing why the data set is no longer available.

PARTICIPANT: I wonder if anyone would like to reflect on the citation systems that have two parties who can be credited: the publishers and data providers. I just noticed that the people we have at this workshop are mostly from the provider side. Are we designing a system that is driven by the data centers and their interests, but not necessarily by the data providers?

DR. CALLAGHAN: Basically, data providers are interested in getting data citation working. We know this because we have asked them, at least in the meteorological sciences. We also have had a few cases of data providers coming to us and asking, "can we get the DOI for our dataset?" or "when will the data be citable?" So, there is interest in the scientific community. As data center managers, our job is to get data from the data providers, but if they do not show any interest in data citation then it is not in our interest to do anything about it.

MR. PARSONS: I will briefly add that if we have the identifier, we do not really need to include the role of the distributor or publisher in the citation. I would like people to think that they are getting a higher quality product out of the National Snow and Ice Data Center, but then I think we also have to be careful about citation being the credit mechanism. For example, I am getting push-back now from the funding agencies that want to have NASA or NOAA in the citation. We never did that with literary citations. Why do we have to do it now?

DR. CHAVAN: I think while it is clear that the basic motivation for data citation is mainly for the authors of the data, publishers of the data could get their work properly recognized as well. If you look at it from the usability perspective, there is equal responsibility on the part of the users of the data by making sure that they cite the data as adequately as possible. This brings in some complexity (e.g., when data are contributed by multiple parties) when the user actually uses records from each of the datasets or subsets of the data. This is exactly why we have been promoting the practice of having user-driven citations on top of the publisher's data citation. I think that it was technically vital previously to have author-driven citations or publisher-driven citations, but I think as the digital area progresses we will need to promote both. The user-driven citations will be the key to authenticate or verify the validity of the interpretations on the dataset that they have actually used.

PARTICIPANT: Is there a possibility that Microsoft can tell me my true worth? If we are already indexing 24 million documents and doing many other things that are not measured in terms of scholarship, we might be able to begin to get at that through the Microsoft Bing index. I could actually come up with a number that was measured across all of my scholarship. This would seem to be something that could change the way people think about how scholarship is measured. It seems we need that kind of metric and maybe it is within reach.

PARTICIPANT: This is the kind of metric that we would like to make available. We are not affiliated with the Bing index, but we have the opportunity to work with that team, combine indexes, and run those kinds of searches and present data in a meaningful way. We are not doing that right now. If the community comes together and indicates that they would be very interested, that would be a good step forward. Maybe there could be a large-scale aggregator service of data citations?

PART FIVE

INSTITUTIONAL PERSPECTIVES

21- Developing Data Attribution and Citation Practices and Standards: An Academic Institution Perspective

Deborah L. Crawford[1]
Drexel University

I have a somewhat unique perspective on this subject. Until September 2010, I worked at the National Science Foundation (NSF), where I was involved in the fashioning of NSF's data management plan policy. Shortly afterwards, I returned to academia, joining Drexel University. I have the pleasure now of implementing the policies that I had a hand in preparing. It is an important topic with a lot of complexity.

Today, I will try to share with you my view from a university administrator's perspective – but I will also touch on the respective roles and responsibilities of academic researchers as individuals and as members of research communities. I was asked to respond to the following question: How are university administrators thinking about data citation and related issues? What follows are some of my thoughts on this subject.

In my role as a vice provost for research at Drexel, I view the stewardship of research data as one of a number of responsibilities I have to create an environment that supports the responsible and ethical conduct of research in the public interest. Developing such an environment has implications for the management of the increasingly digital research data that we collect or create.

Let me first talk about the role of researchers, and the research communities to which they belong, in data stewardship. As is already quite well known, there are significant differences in practices among scientific communities, including the communities represented here at this workshop.

For example, some of our communities have, for a decade or more now, leveraged the economies afforded by data sharing, attribution, and citation. These tend to be the scientific and engineering communities, where data have been and continue to be created or collected with the intent to be shared broadly. These include, for example, environmental and astronomical sciences, and geosciences communities—typically those communities where data are collected on nationally or internationally-supported and community-governed instruments or facilities. And now, thanks to the "omics" revolution, a number of the life sciences communities too are generating data with intent to be shared.

In other fields, cultures continue to be much more individual investigator oriented. In such domains, the independence of individual investigators is fiercely guarded and research data are rarely shared, except in relatively modest ways through peer review publications. I think it is useful to keep these cultural and research differences in mind as we think about how we move forward–one size is unlikely to fit all.

---

[1] Presentation slides are available at http://sites.nationalacademies.org/PGA/brdi/PGA_064019.

We need to develop explicit policies on data sharing, attribution, and citation–both domain-based policies for the scientific and engineering communities, and institutional policies that complement and support community policies. It is important that we develop these policies and supporting practices in a collaborative way, bringing all stakeholder groups along so that we can fully leverage the added value of the enormous and growing quantities of digital data to advances in science and technology.

Let us now turn our attention to the role of academic institutions. Just as some communities have well developed data policies and practices and others do not, so some institutions have data sharing policies and others do not—at least, not yet.

In tenure and promotion policies and practices that pertain to data sharing, citation and attribution, culture matters very much. For investigators in communities more accustomed to data sharing, data attribution and citation is likely to be valued in tenure and promotion decisions. This, however, is not true across the board. So in fashioning academic policies that promote data sharing, citation and attribution, we must be mindful of, and manage for, these differences.

Institutions should help faculty understand what is expected of them in the responsible stewardship of research data in our increasingly digital scientific world. Deans and department heads are major institutional stakeholders too, for they must provide leadership in raising awareness about this important topic and its implications in matters such as tenure and promotion (and others), and they must serve as advocates for change, where necessary.

I believe mid-career faculty play a very important role. We cannot expect our junior faculty, who are often pioneers or early adopters of new digital research modalities, to carry the weight of promoting the development of new data policies, for they have too many other pressures coming to bear on them, and in fact might be penalized for having pioneering views. Mid-career faculty members are likely to be key to moving a conversation forward on these topics. They are the ones who typically are more engaged in research where progress demands an increasing reliance on the sharing and attribution of digital data, and these faculty members may be more willing and able to speak to and be heard about the importance of these issues.

Let me now briefly address the issue of institutional repositories. Many of us believed that institutional repositories, interoperable ones of course, would be a key to the future; they would enable universities to actively manage their digital assets, manage their intellectual property with appropriate controls, and explore new forms of scholarly communications. The role of institutional repositories is especially important in the later stages of the data lifecycle, as researchers focus on new and interesting scientific opportunities and worry less about the research data of their past interests. Thus, institutional repositories were expected to play important roles in data curation and preservation.

In practice, however, institutional repositories are not living up to our expectations, partly because researchers are not routinely depositing their digital objects in the repositories that their institutions are providing. Many researchers do not see the value to their science and to their careers of doing so, which ultimately is the bottom line for most researchers. This is something we need to keep in mind as we think about the ways in which institutions encourage faculty to engage in conversations about the future of data sharing, citation, and attribution.

It is important to note that it is not at all clear that academic institutions across the board are in a position to move boldly into this new world. For one thing, as we heard yesterday, universities have not been significantly engaged in the active long-term management of research data to date. Traditionally, the majority of investigators have managed or have been responsible stewards of their own data, where community governance of data was essential to advances in certain fields. It is fair to say that there is much more evidence of community-based initiatives, albeit in some fields more than others, than there are university-level initiatives.

Equally important, or maybe more important today, the substantial cost implications of providing long-term stewardship of data is a very significant concern for research universities. In an increasingly difficult economic environment, where concerns already exist about the escalating costs of higher education and where the federal government is unable or unwilling to support the full cost of research in the academy, who bears the responsibility for paying to ensure long-term open, useful access to research data created in the public interest? This is a policy issue that the government-university partnership needs to resolve.

The discussions that we have been having in this workshop and in recent years raise important questions about who is or should be the champion for these issues at the university, and who should do what? Unfortunately, there are no clear answers at most universities because of the complexities. Faculty, researchers, and students have a voice and a role, but for the most part, they are not substantially involved in conversations because, for the most part, the value to their science and to their careers is not readily apparent to them. Deans, colleges, and schools have voices as well, but for the most part, they are not serving as change agents. This is true, too, in university research offices, in part because they tend to reflect the cultures of the research communities they represent, and undoubtedly, because of concerns about costs. Libraries have served as the strongest advocates for these kinds of changes, but they probably do not have the institutional power or authority to really effect change. So, if we are going to make a difference, there has to be a clear change mandate in institutions involving all institutional stakeholders.

This is a critical topic of conversation for academic institutions, because it impinges upon their reputations as essential contributors to the national knowledge enterprise.

22- Data Citation and Data Attribution: A View from the Data Center Perspective

Bruce E. Wilson[1]
Oak Ridge National Laboratory and the University of Tennessee

One of the several things I am doing right now is pushing the question of how to enable scientific collaboration and, in particular, how to put in place the technology to do that. I am also working with my colleges at the University of Tennessee to look at these issues from a sociological perspective.

For three years, I was the manager of the Oak Ridge National Laboratory (ORNL) Distributed Active Archive Center (DAAC) for biogeochemical dynamics. It is one of NASA's Earth Observing System Data and Information System (EOSDIS) data centers managed by the Earth Science Data and Information System (ESDIS) Project. The ORNL DAAC archives data produced by NASA's Terrestrial Ecology Program, as well as data of particular interest to scientists funded by that program. The ORNL DAAC provides data and information relevant to biogeochemical dynamics, ecological data, and environmental processes that are critical for understanding the dynamics relating to the biological, geological, and chemical components of Earth's environment.

I also spent eighteen years in private industry. I mention that because this experience influences in some ways my perspective on data citation and attribution issues. In addition, I have had some involvement in projects with the National Biological Information Infrastructure, at the USGS, and I continue to work on the citizen science side of data submission and data citation for the U.S.A. National Phenology Network (USA-NPN).

At the ORNL DAAC, we make sure that the data generators get credit for what they have done. An incentive for data attribution is to ensure that the data center gets credit for hosting the data as well. We also want to understand what data are or are not being used. For example, we have good statistics about how many times datasets were downloaded. These data show that, on average, it takes about 18 to 24 months from when a dataset gets downloaded to when it gets cited, if it gets cited or referenced at all. We found that some datasets are downloaded more, but may be cited less. Does that indicate that the data are hard to use, that there are barriers to being used, or that there is something wrong with the data? There is typically no single answer to that question, and it is something to investigate on a dataset-by-dataset basis.

This kind of use metrics (not only of downloading, but of actually using a dataset) can be extremely valuable for the data center to understand its business model and operational environment. For example, it helps us understand how the data are being used outside of the scholarly context.

Tracking the use of an individual dataset is important, but what is the value for a data center if these data were used in a university study, for example? What is the value to the data center and

---

[1] Presentation slides are available at http://sites.nationalacademies.org/PGA/brdi/PGA_064019.

to the sponsor if we did something that made the data easier to use in undergraduate education classes? So, it is arguably valuable to the science and scientific community, but how do we measure that? How to we understand those kinds of uses?

Here is an example of dataset citation:

> Gu J. J., E. A. Smith, and H. J. Cooper. 2006. LBA-ECO CD-07 GOES-8 L3 Gridded Surface Radiation and Rain Rate for Amazonia: 1999. Data set. Available on-line [http://www.daac.ornl.gov] from Oak Ridge National Laboratory Distributed Active Archive Center, Oak Ridge, Tennessee, U.S.A. doi:10.3334/ORNLDAAC/831.

We started adding the DOI to this citation style about five years ago. The major reason for adding the DOI was that we had a citation without the DOI and some journals rejected it because it was not a valid long-term citation.

We put the DOI into it because the DOI is an established standard and one that the publishing community, in particular, uses. Using the standard that makes sense to publishers helped to reduce the barriers to adoption of data citations. And a second key point is that the citation contains key information we need: the names of those who created the dataset, it tells where to find the data, and it has a persistent identifier

Let me now focus on the data center roles and responsibilities in this process. I think a key role here is the issue of stability. Data centers need to provide stability through using persistent identifiers and ensuring technical, social, and organizational sustainability.

Data centers also have to encourage use and make it easy for users. We need to make it easy to download datasets like a bibliographic set. Furthermore, we have to work on challenges such as what does the identifier points to, how to handle continuous data, sub-setting, on-demand data, as well as issues related to scalability of the data management process. We have to work on some fundamentally different paradigms and be willing to take some risks about changing our business model.

Let us now see how our datasets actually get used. The Figure 22-1 shows the growth in cited versus referred databases. Compiling the information for this table was an effort of multiple weeks by the ORNL Library and the ORNL DAAC staff.

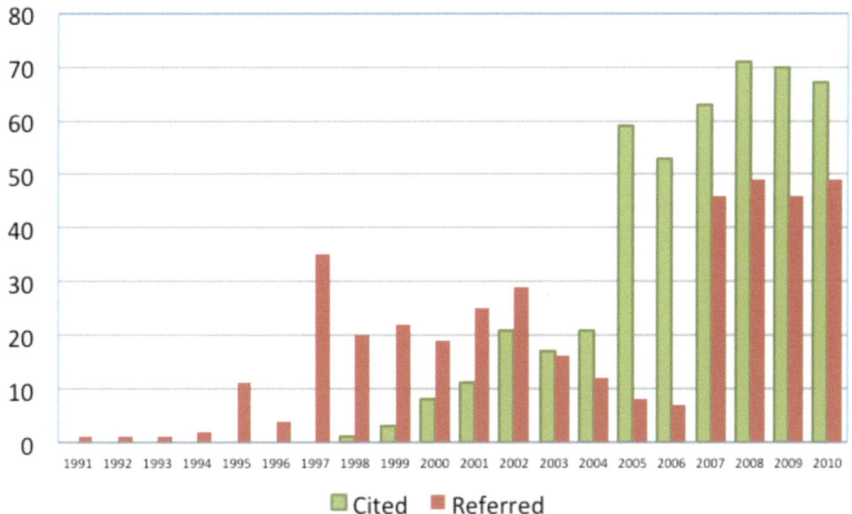

FIGURE 22-1 Growth in cited vs. referred databases.

This diagram shows an increased number of datasets cited versus referred. By "cited" we mean that the ORNL DAAC dataset is in the bibliography or in the requested format; and by "Referred" we mean that we infer, from the text of the article, that one or more ORNL DAAC datasets were used in the work. ORNL DAAC requests that data be cited in the list of references.

This time series provides evidence of some changes in scientists' behavior towards data citation and attribution. The best way to facilitate and promote this change is through having more champions. Academicians and scientists are only going to do something if somebody who is successful tells them that it has worked.

23- Roles for Libraries in Data Citation

Michael Witt[1]
Purdue University

As a practicing librarian, I will be focusing on the roles for librarians and information professionals in data citation and attribution. I would like to start by answering the question, Why are librarians involved in data, and why are they interested in data citation? If we go back to the workshop on "New Collaborative Relationships: The Role of Academic Libraries in the Digital Data Universe" that was sponsored by the Association For Research Libraries (ARL) and the National Science Foundation (NSF) in September 2006, an important need was identified "…for new partnerships and collaborations among domain scientists, librarians, and data scientists to better manage digital data collections; necessary infrastructure development to support digital data; and the need for sustainable economic models to support long-term stewardship of scientific and engineering digital data for the nation's cyberinfrastructure."[2]

To follow up, in August 2010, the ARL did a survey of its member institutions (approximately 130) and 57 of them responded. Some of the findings include: (1) 21 of them currently provide infrastructure and services for *e*-Science and data support, and (2) 23 members are in the planning stages.[3]

This shows that libraries are involved in this area of data curation, at least in the context of academic and research libraries. That is not to say that any of these issues are exclusive to those libraries. In fact, I think that a lot of these needs extend to public libraries and citizen science, and other libraries outside of the university context.

I propose that data citation has "a last mile problem." In communication networks it is usually easier to connect countries and cities than it is to connect to individual end-nodes, such as houses, especially in rural areas. In the data citation arena, the challenge is: how do we reach and affect a change in practice among end-users of data? How can we reach people who will be writing papers and citing the data? Those users could be students, faculty researchers, citizens, or government agencies, etc.

I believe that a role that librarians can play here is rooted in libraries' tradition of information literacy outreach and instruction. Information literacy is a set of abilities requiring individuals to recognize when information is needed and have the ability to locate, evaluate, and use effectively the needed information.[4] This includes the proper citation and attribution of sources.

---

[1] Presentation slides are available at http://sites.nationalacademies.org/PGA/brdi/PGA_064019.
[2] Available at: http://www.arl.org/pp/access/nsfworkshop.shtml.
[3] C. Soehner, C. Steeves, and J. Ward, E-Science and Data Support Services: A Study of ARL Member Institutions Association of Research Libraries, 2010. http://www.arl.org/bm~doc/escience_report2010.pdf.
[4] American Library Association. 1989, Presidential Committee on Information Literacy. Final Report.

If you look at the Information Literacy Competency Standards for Higher Education[5] from the Association of College and Research Libraries (ARCL), you can replace the word "information" with "data" and the competencies make sense and remain relevant.

Where can users look for information on how to cite data? One natural place to turn would be style guides. I did a study with two colleagues, where we looked at 20 different style guides and performed content analysis to see what kind of instructions they are providing users explicitly to cite digital data. The answer is: they do not consistently address data citation and attribution.

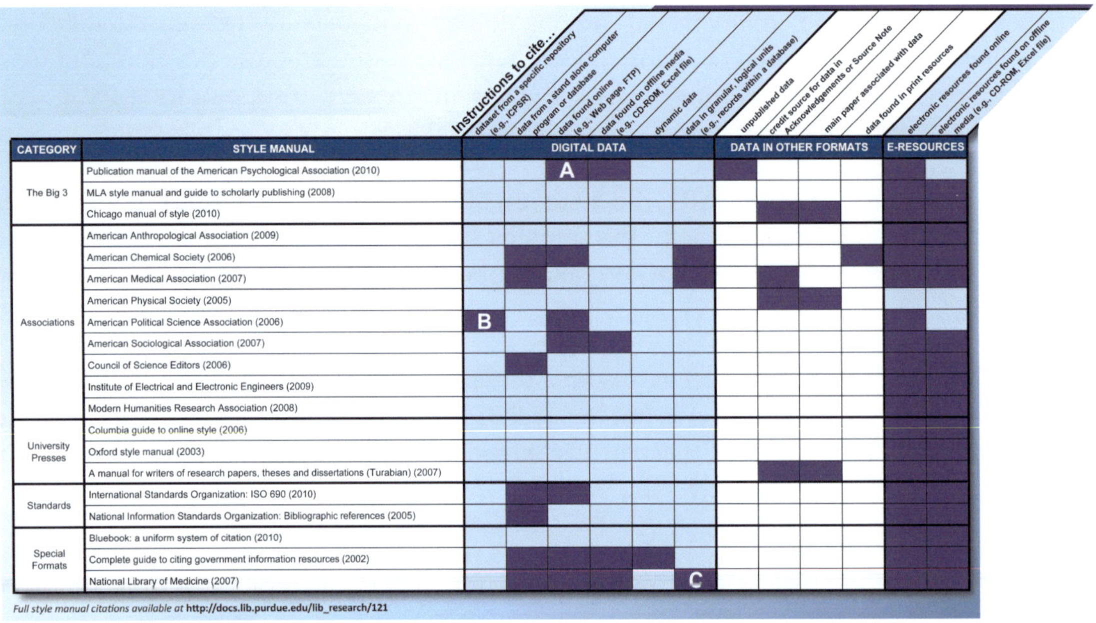

FIGURE 24-1 A Description of Data Citation Instructions in Style Guides.
SOURCE: International Digital Curation Conference, Chicago, IL. Retrieved from http://docs.lib.purdue.edu/lib_research/121/. Newton, Mooney, & Witt. (2010).

If you look at the above grid, it covers instructions for digital data, data in other formats (e.g., paper-based tables), and other electronic resources. The dark purple indicates the areas where the style guide provides explicit instructions for citation. The light colors (i.e., aqua or white) indicate that there are no explicit instructions. So, generally speaking, some style guides do a better job than others—but if this is where students and others are turning for instructions to properly cite data, they will undoubtedly be frustrated.

---

[5] Available at: http://www.ala.org/ala/mgrps/divs/acrl/standards/informationliteracycompetency.cfm.

One thing that we see happening on our university campuses is that librarians are stepping in to address this need by creating resource guides. This is a common practice of librarians to develop bibliographies and path-finders to introduce topics and tools to users. Here are some examples of resource guides on data citation that are appearing at universities from their libraries:

- MIT: http://libraries.mit.edu/guides/subjects/data/access/citing.html
- MSU: http://libguides.lib.msu.edu/citedata
- Minnesota: http://www.lib.umn.edu/datamanagement/cite
- Purdue: http://guides.lib.purdue.edu/datacitation
- Oregon: http://libweb.uoregon.edu/datamanagement/citingdata.html
- Cambridge: http://www.lib.cam.ac.uk/dataman/pages/citations.html
- Virginia: http://www2.lib.virginia.edu/brown/data/citing.html

These guides are written by librarians in most cases and tailored for their particular audience. They may be tailored for undergraduate or graduate students, faculty researchers, or others.

One project that I would like to briefly talk about is Databib.[6] This project was funded through the Institute for Museum and Library Services (IMLS). Here is the description of the project:

> The libraries of Purdue University and Penn State University will partner to create a new online information resource for research data producers, users, publishers, librarians, and funding agencies. This resource, Databib, will be an annotated online bibliography of research data repositories, created and maintained by an online community of librarians. Databib will be an important focal point for connecting librarians more closely with other research data stakeholders and demonstrating the significant contributions libraries can make to solving the challenges posed by digital datasets. The Databib platform will also serve as a testbed for linking, integrating, and presenting information about datasets in new ways.[7]

Databib is essentially a bibliography that describes data repositories. What we are doing is creating a platform for librarians to submit and enhance bibliographic entries that describe these data repositories and do it in a way that is maximally open, using the Creative Commons Zero public domain protocol. If someone wants the list or the metadata, they are free to download and use them. Also, if someone wants to enhance the metadata or annotate them, that is also possible.

We are creating this resource for the community to help users find data as well as to help data producers identify repositories where they can submit their data, to share this information with funding agencies that mandate data management and tell them where data have been submitted, because these directions are unclear in many cases. We also want to test the notion of a bibliography. We will have bibliographic records that can be exported as MARC records, so if someone wants to download them into their library catalogue, they can. Also, if someone wants to integrate them with other Web 2.0 tools, such as social tagging and social bookmarking, Databib will facilitate sharing links and citations. Finally, we want to use this platform to experiment with linked data. We want to create a platform where the descriptions of these data

---

[6] Databib website, http://databib.lib.purdue.edu.
[7] IMLS press release, http://www.imls.gov/grant_awards_announcement_sparks_ignition_grants.aspx.

repositories can be linked in as many ways as possible to other things, whether it is in the same subject area, same agency that supports the data repository, or any other level of linkage. This project is a nine-month project, and Databib will be going online in the spring of 2012.

Going back to the potential role of libraries and librarians, libraries are a primary actor in the scholarly communication chain. I believe that libraries can promote persistence for links to data. Jan Brase talked about DataCite yesterday. There are many libraries that are participating in this effort. I think that libraries need to adopt URI policies. We are creating a lot of digital content and making it available in a lot of different ways with links that break. So, in addition to minting and maintaining unique, global, and persistent identifiers, we can have more general URI policies, which we can advocate for web content across our institutions.

Are libraries presenting our own data in ways that facilitate or encourage citation? Libraries maintain institutional repositories and other digital libraries where they are presenting digital objects, but do we have supporting documentation and FAQs that give users instructions for citation? Do we provide embedded, structured metadata within the web page, such as COinS, micro-formats, or RDF? Do we facilitate exportable citations? Many of our libraries have data services that are doing outreach to faculty members to help them understand data management plans. Before projects are funded and data are generated, there is the opportunity to have a conversation about data-sharing with the different stakeholders. There is an opportunity for advocacy.

I would like to raise awareness of the work being done by the International Association of Social Science Information Services and Technology (IASSIST). I co-chair a special interest group on data citation with Mary Vardigan. Among the over 300 members that IASSIST has, about 40 or 50 of them are involved in this special interest group. Some of the activities that we have been engaged in include an effort to derive a common set of user instructions for citing data. We realize that we would not necessarily be able to use a perfect set of instructions for all cases, but if we can come up with a core set of instructions, that would be very useful. Also, there has been some work to integrate datasets as a resource type in citation management software such as EndNote or RefWorks. Moreover, we are doing some advocacy. We have been writing letters to style guides editors and publishers to encourage them to articulate policies and instructions for data citation to their authors. Also, like many other special interest groups, we are generating resources such as a website and brochure that are publicly available for use.

To conclude, librarians and information professionals can play important roles in advocacy and outreach, and in the integration and citation of data. This includes data citation in reference services and information literacy instruction and standards. Librarians should ask themselves: if we are publishing data, are we making our data citable, and are we incorporating data into information literacy?

One last observation: many libraries are creating new data services units that can help raise awareness of and address issues related to data attribution and citation for their communities. Promoting proper data use and citation should be a part of what we normally do in libraries, a part of our regular practices. There seems to be a trend of libraries addressing research data in a specialized manner, e.g., "data reference" and "data information literacy". I suggest that, after a period of time, the library profession will become more comfortable with data and will not need to qualify "data" services as such. The same principles of library science that apply to traditional formats can be applied to data.

The timing seems to be perfect for people to connect and collaborate to address data citation and attribution issues.

24- Linking Data to Publications: Towards the Execution of Papers

Anita De Waard[1]
Elsevier Labs and the University of Utrecht, The Netherlands

First, I would like to say that I am not representing all commercial publishers and that I have not even coordinated this talk with my colleagues at Elsevier, so this is my personal perspective on the issues being discussed here.

I think it is useful when we are talking about integrating data with publications to look at where data fit within the scientific process. The KEfED model developed by Gully Burns[2] can help in this regard.

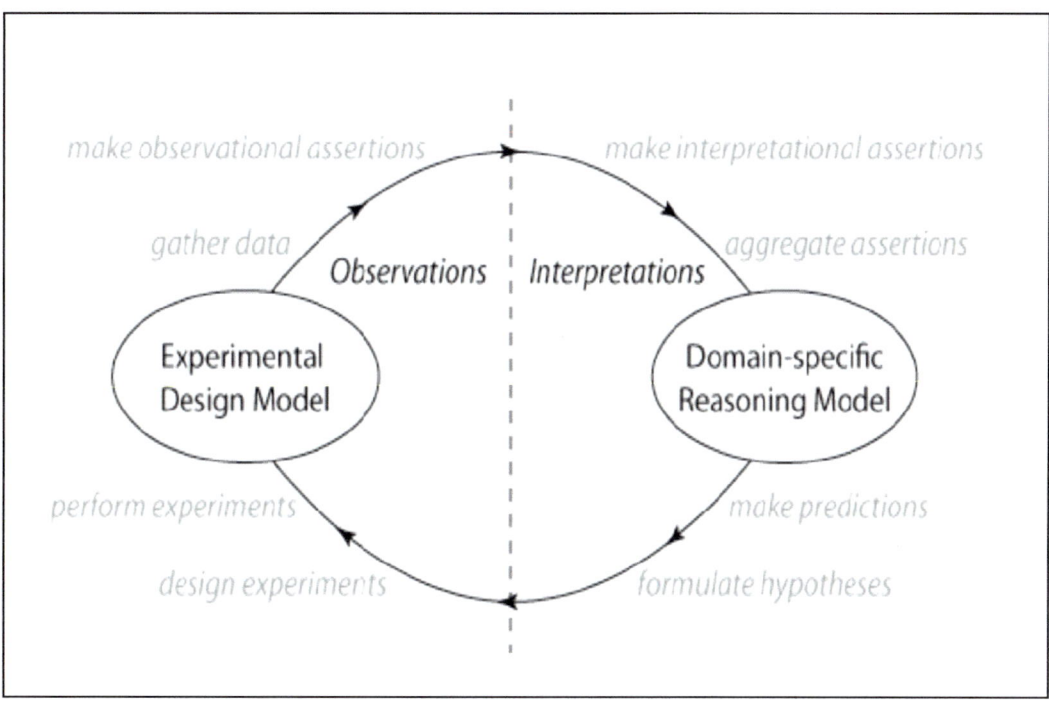

FIGURE 24-1 KEfED model "Cycle of Scientific Investigation."

Essentially, in doing research we start thinking about the background and making some hypotheses. This is basically experimental science. You do an experimental design, you manipulate some external objects, and then you have observations. From those observations, you gather what is called data. Then you do some statistical analysis, and come up with some findings. In general, the data support your claims and findings. What happens in a publication is

---

[1] Presentation slides are available at http://sites.nationalacademies.org/PGA/brdi/PGA_064019.
[2] Gully APC Burns and Thomas A. Russ. 2009. Biomedical knowledge engineering tools based on experimental design: a case study based on neuroanatomical tract-tracing experiments. In *Proceedings of the fifth international conference on Knowledge capture* (K-CAP '09). ACM, New York, NY, USA, 173-174. DOI=10.1145/1597735.1597768 http://doi.acm.org/10.1145/1597735.1597768.

that you make a representation of your thoughts through language. These are the bases with which I would like to start.

Currently, the scientific community is storing data in repositories. We link to publications and vice-versa. The example that is commonly used is that people add PDFs and spreadsheets to their papers. This is pretty useless because we are not doing anything with these documents. Having them does not mean we can find the dataset.

In general, I believe that datasets should all be available for server search and that sets and subsets of that data should be made freely accessible, whenever possible. Overall, commercial publishers are not interested in owning or charging for research data or running those repositories. There might be exceptions, but in general, this is the case.

In my view, most publishers are very interested in working with data repositories and believe that it would be very useful if there were one place where we can find data items. It would be useful if an identifier is persistent and unique and that if the content changes, the identifier changes as well. Also, it would be very useful if the data would link back to the publication. It would be more interesting if we have data in a repository and can link them to some content from within a publication. Not only from the top level, but from within the publication. There are some examples of this. What my lab has been doing currently is tagging entities and linking them to databases. This involves some manual as well as some automated work.

More interesting, I think, is the fact that we can now create claim evidence networks that span across documents, so we can have a statement that can be backed up in a table or a reference in another publication or in another data center. At least at Elsevier, we are very invested in the idea of linked data. We have developed something that we call a satellite, which is essentially a way to describe a Linked Data annotation, in RDF. We are using Dublin Core and SWAN's provenance and authoring/version ontology to identify the provenance.

We are very happy to develop this with people like Paul Groth and Herbert van de Sompel and others to have an ontology that connects to their work. The idea is that we can have some files that link to our XML at any level of granularity. There are files that sit outside the publication or the data center but we can still link one to the other. I think this is a very promising way to move forward.

What would be really interesting is if we had the opportunity to completely re-think science publishing. Why only change where the data is located: why not change the whole process? In my opinion, what is key is that scientists should be allowed to do their research process the way they want. We do not want to put more obstacles in front of the busy scientists who are already struggling to do their work. In fact, I think that the publishers would like to help them. So, if they have an experimental design, perhaps they can put a copy of it in the repository and put a link to it in their paper. Similarly, there are reports of observations. Perhaps there can be some way to deposit these reports in a repository and to pull them into their paper, code their statistics in a same way, and then draw the conclusions.

For the publisher and probably for the reader, it is incredibly important to maintain the context that the data have (e.g., the experimental context, the reason you did the experiment, the time involved, and the like). There is a narrative context and we are using it to prove a point, so the data act as a key point for life scientists to communicate with other scientists. There are big questions that we are tackling and it is very important that the data are maintained and preserved.

Now let me ask this question: why do not the scientists themselves keep track of their own experimental design, their observed results and their code of statistics? They can share part of this with the publisher. Similarly, they can share with the data repositories. They can share the experimental design, the data and the code of statistics, using cloud computing. Imagine scientists using the cloud to store their research, find their results, experiments, and observations. I think it is truly important that as research keeps building, there are good systems in which researchers can keep track of their own data, store them, and add appropriate metadata.

The assignment of unique identifiers plays a central role in the advertisement of these materials. Data centers are able to connect datasets and promote them. They can also advertise them. The role of data centers in terms of quality control and access is very critical and, as we saw earlier in this meeting, this differs from one field to another.

So, if we are publishing a paper with data, all we need to do is to deposit our document in a repository and allow access to an editor or somebody who we think can evaluate our work. Then, we would have access to the collective thoughts as well as to links to the data, to the workflow, to the other science components, and to a publisher or somebody in the role of validating quality.

I think these and similar practices will connect more in the future and publishers, data repositories, and perhaps software developers (e.g., Microsoft, Google, Skype, Twitter, or Dropbox) will be involved in these processes. We all use commercial software all the time. These programs are very good at building tools that help us communicate. Therefore, it is very useful to have such companies working with us on improving communication between scientists by encouraging them to build better software and applications.

Citizen science was mentioned earlier as well. Citizens can also play a key role in these processes and we should be keen to involve them. Again, some technological components and applications are now in place and can facilitate these processes.

Let me conclude by emphasizing that, in my view, publishers are not interested in owning or charging for data. We believe in identifiers and embrace open standards and I think that scientists should keep track of their own work. We certainly believe in a future where science is shared and stored in a better and productive way, as well as in working together with all stakeholders to make it happen.

25- Linking, Finding, and Citing Data in Astronomy

Michael J. Kurtz[1]
Smithsonian Astrophysical Observatory

My presentation is focused on data citation and attribution issues in the field of astronomy. There are commercial astronomy journals, but most of them are not very important. Basically, the entire system is operated through collaboration between data centers and publishers, where the publishers are the professional societies. I thought I would first share with you some information about a similar workshop held 25 years ago. This Astrophysics Data System (ADS) workshop (1,2) was held on 1987 to discuss issues related to:

1. Data Accessibility
2. Data Format Standards and Quality
3. Data Analysis and Reduction Software
4. User Scenarios
5. Observation Planning and Operations

There was a report from this workshop. One of the points that the report made was the following:

> There is an urgent need for a master directory for all NASA space-based observations. It is recommended that the directory should include all past observations and currently planned observations from observatories, and the past and planned observations from ground-based observatories, where possible. NASA and NSF should enter into discussions regarding how this can be accomplished.

If we think about what we are trying to do in this workshop, it is basically similar to this 1987 workshop. If we make a list of all the observations, give them names and addresses, that is essentially putting DOIs and addresses on every piece of data. However, this has never happened. They spent over the next seven years or so $25 million-$30 million trying to make this happen. The reason it did not happen was primarily control. None of the archival systems were willing to give up the control necessary to make it happen. Now, it is a quarter of a century later and that still is the case.

The second issue that I would like to talk about is related to the American Astronomical Society's (AAS) policy for dataset linking. Their policy started seven or eight years ago, is active and people are joining up. The big data centers can create tags or names for their datasets. They are able to do it the way they want. There is no large system telling them how to do it. However, they have to agree that they will be able to resolve these tags with the datasets when asked. This is a great effort and all the big U.S. data centers are part of it, but the Europeans did not join.

We now will look at data citation and attribution in practice. We have been doing this for a couple of decades now. It pretty much works and it is growing organically. Below is a 17-year old image. It is the first image of a web browser used in our ADS system.

---

[1] Presentation slides are available at http://sites.nationalacademies.org/PGA/brdi/PGA_064019.

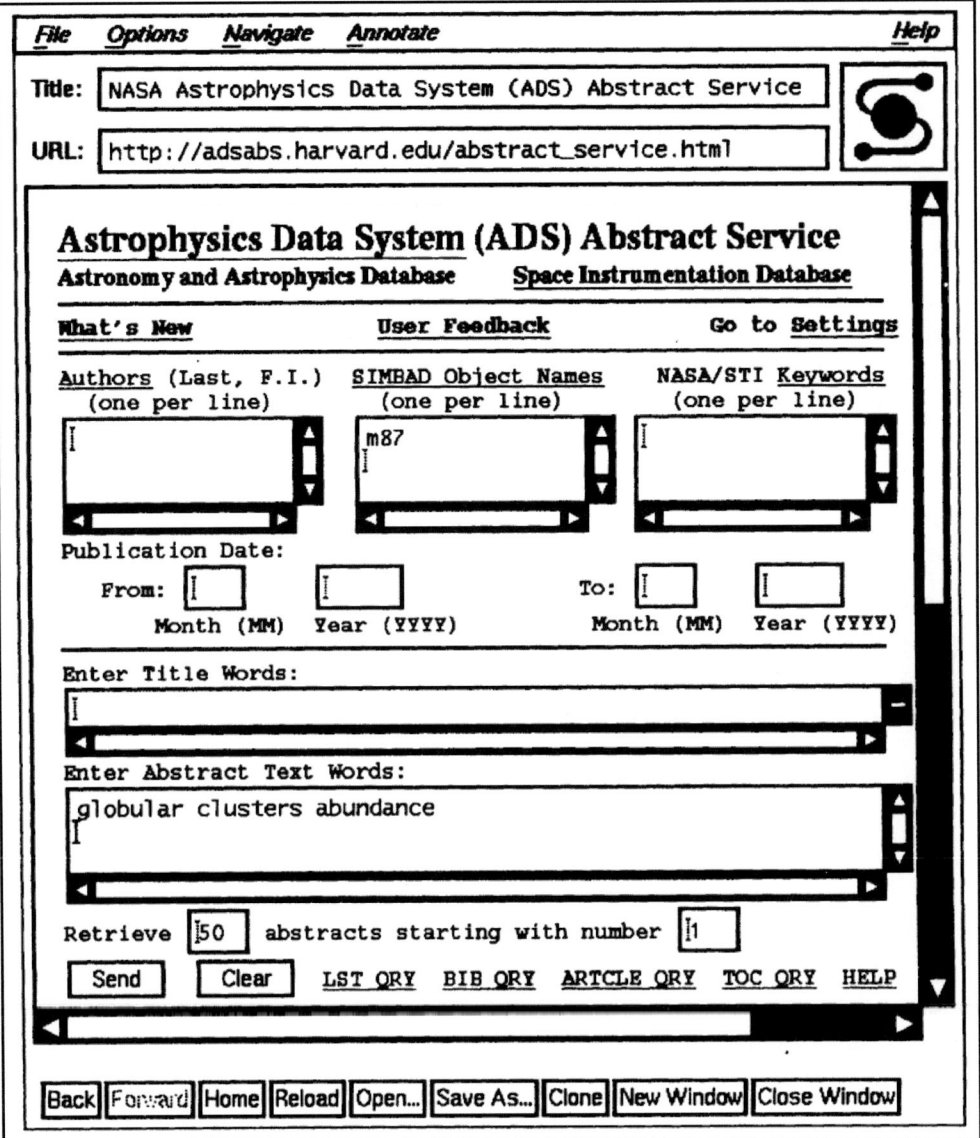

FIGURE 25-1 First web browser used in ADS system.

What you can see at the bottom is a complex literature query. To an astronomer, "abundance" means the fraction of different elements in a star. It is also called metallicity. The next image is what it looks like today. It is basically the same query, except that we have changed the word abundance to metallicity.

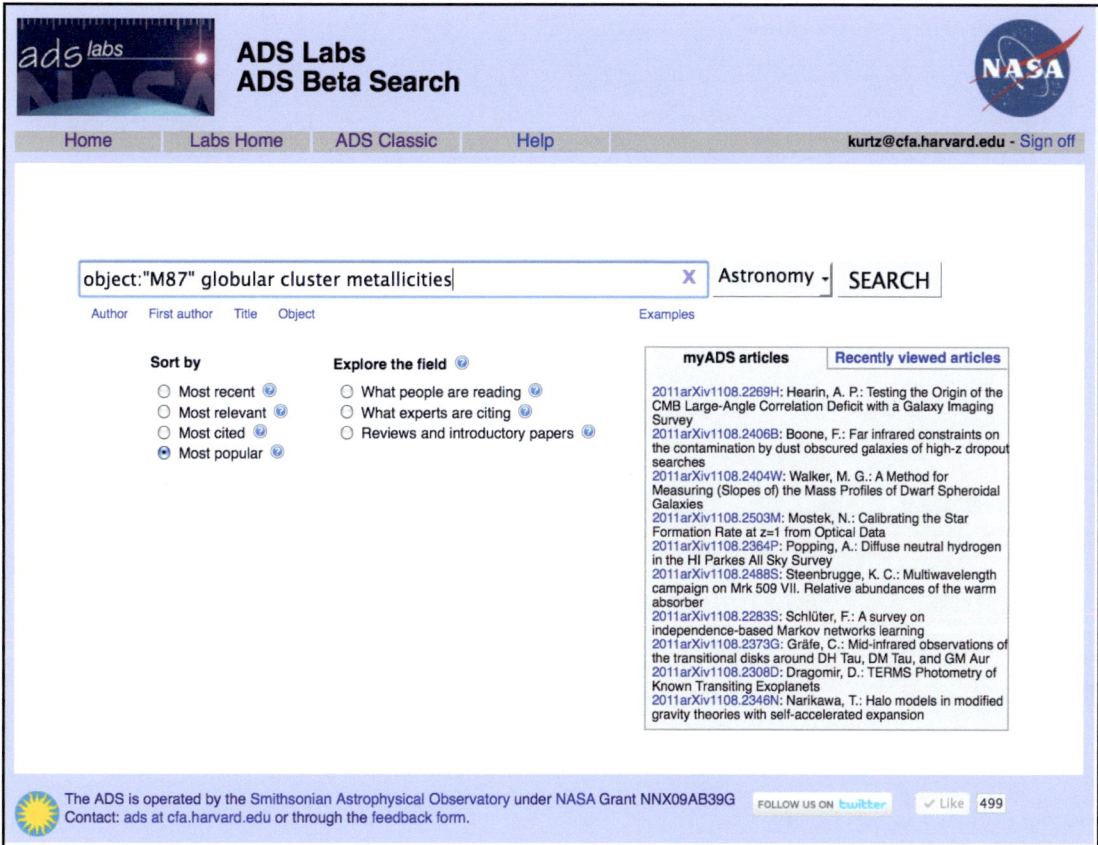

FIGURE 25-2 Complex literature query.

What happens when you run a query is that a list of popular papers about the metallicity of M87 comes up. We are interested in data so we can ask the system to select only papers with links to data from the space telescope; this yields a list of seven papers concerning the metallicity of the galaxy M87 which have links to on-line data in the HST archive.

We could just as well have chosen any (or all) of about a dozen other archives to obtain original (from the telescope) data, or chosen to retrieve tabular data from any of several dozen papers.

The next image is from Todd Vision of Dryad. It shows the two different kinds of data to which the ADS is linked.

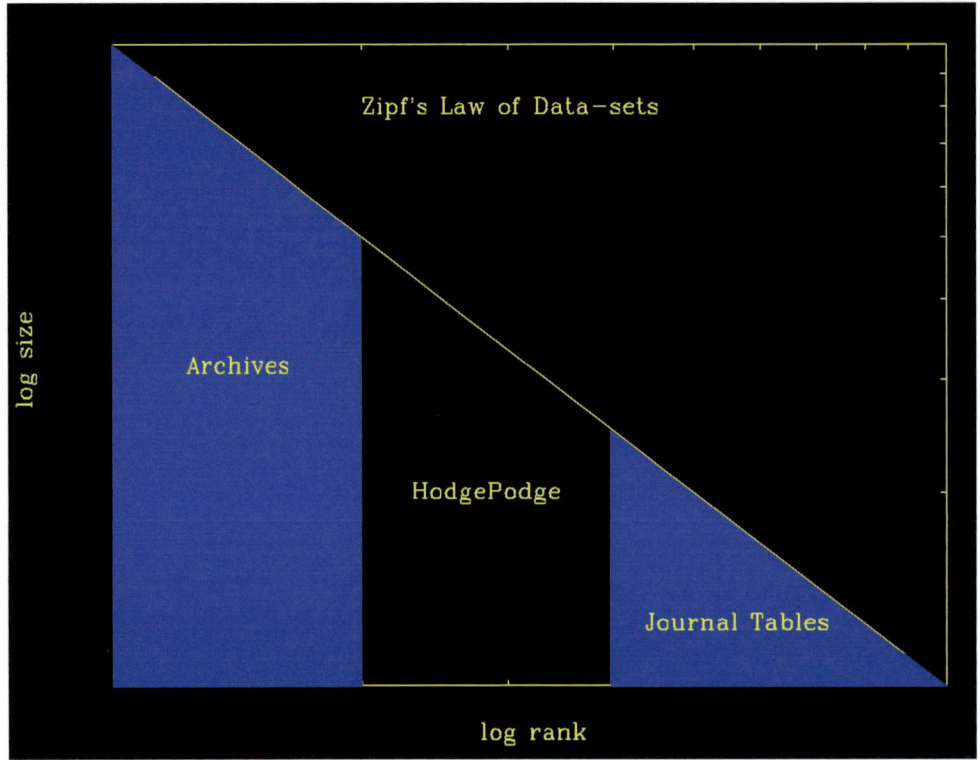

FIGURE 25-3 Zipf's Law of Datasets
SOURCE: Todd Vision, Dryad.

This figure shows that the highest-ranking source in terms of size are the archives. The archives are enormous in the amount of data they hold and most of them are very well managed. They have people who curate their data and make links. Some of these archives are better than others but in general, they are good. On the right-hand side, there are small data tables from journals. There is a system for taking these tables numerically and keeping them online. Most of these functionalities have been running pretty smoothly for about 20 years. Finally, the middle part is where the problem is. There we find small and medium size datasets with no home; they are often too big for the journals, but are not part of the established archives.

I will conclude by talking about a paper that Margaret Geller and I submitted to the *Astronomical Journal* last week. I am going to show you the references for a plot that we made in the paper. The first one is from the Sloan Digital Sky Survey (SDSS). It shows that we used York, *et al.*, but we did not use the paper, we used the database.

THE ASTRONOMICAL JOURNAL, 120:1579–1587, 2000 September

THE SLOAN DIGITAL SKY SURVEY: TECHNICAL SUMMARY

DONALD G. YORK,[1] J. ADELMAN,[2] JOHN E. ANDERSON, JR.,[2] SCOTT F. ANDERSON,[3] JAMES ANNIS,[2] NETA A. BAHCALL,[4] J. A. BAKKEN,[2] ROBERT BARKHOUSER,[5] STEVEN BASTIAN,[2] EILEEN BERMAN,[2] WILLIAM N. BOROSKI,[2] STEVE BRACKER,[2] CHARLIE BRIEGEL,[2] JOHN W. BRIGGS,[6] J. BRINKMANN,[7] ROBERT BRUNNER,[8] SCOTT BURLES,[1] LARRY CAREY,[3] MICHAEL A. CARR,[4] FRANCISCO J. CASTANDER,[1,9] BING CHEN,[5] PATRICK L. COLESTOCK,[2] A. J. CONNOLLY,[10] J. H. CROCKER,[5] ISTVÁN CSABAI,[5,11] PAUL C. CZARAPATA,[2] JOHN ERIC DAVIS,[7] MAMORU DOI,[12] TOM DOMBECK,[1] DANIEL EISENSTEIN,[1,13,14] NANCY ELLMAN,[15] BRIAN R. ELMS,[4,16] MICHAEL L. EVANS,[3] XIAOHUI FAN,[4] GLENN R. FEDERWITZ,[2] LARRY FISCELLI,[1] SCOTT FRIEDMAN,[5] JOSHUA A. FRIEMAN,[1,2] MASATAKA FUKUGITA,[13,17] BRUCE GILLESPIE,[7] JAMES E. GUNN,[4] VIJAY K. GURBANI,[2] ERNST DE HAAS,[4] MERLE HALDEMAN,[2] FREDERICK H. HARRIS,[18] J. HAYES,[7] TIMOTHY M. HECKMAN,[5] G. S. HENNESSY,[19] ROBERT B. HINDSLEY,[20] SCOTT HOLM,[2] DONALD J. HOLMGREN,[2] CHI-HAO HUANG,[2] CHARLES HULL,[21] DON HUSBY,[2] SHIN-ICHI ICHIKAWA,[16] TAKASHI ICHIKAWA,[22] ŽELJKO IVEZIĆ,[4] STEPHEN KENT,[2] RITA S. J. KIM,[4] E. KINNEY,[7] MARK KLAENE,[7] A. N. KLEINMAN,[7] S. KLEINMAN,[7] G. R. KNAPP,[4] JOHN KORIENEK,[2] RICHARD G. KRON,[1,2] PETER Z. KUNSZT,[5] D. Q. LAMB,[1] B. LEE,[2] R. FRENCH LEGER,[3] SIRILUK LIMMONGKOL,[3] CARL LINDENMEYER,[2] DANIEL C. LONG,[7] CRAIG LOOMIS,[7] JON LOVEDAY,[1] RICH LUCINIO,[7] ROBERT H. LUPTON,[4] BRYAN MACKINNON,[2,23] EDWARD J. MANNERY,[3] P. M. MANTSCH,[2] BRUCE MARGON,[3] PEREGRINE MCGEHEE,[24] TIMOTHY A. MCKAY,[25] AVERY MEIKSIN,[26] ARONNE MERELLI,[27] DAVID G. MONET,[18] JEFFREY A. MUNN,[18] VIJAY K. NARAYANAN,[4] THOMAS NASH,[2] ERIC NEILSEN,[5] RICH NESWOLD,[2] HEIDI JO NEWBERG,[2,28] R. C. NICHOL,[27] TOM NICINSKI,[2,29] MARIO NONINO,[30] NORIO OKADA,[16] SADANORI OKAMURA,[12] JEREMIAH P. OSTRIKER,[4] RUSSELL OWEN,[3] A. GEORGE PAULS,[4] JOHN PEOPLES,[2] R. L. PETERSON,[2] DONALD PETRAVICK,[2] JEFFREY R. PIER,[18] ADRIAN POPE,[27] RUTH PORDES,[2] ANGELA PROSAPIO,[2] RON RECHENMACHER,[2] THOMAS R. QUINN,[3] GORDON T. RICHARDS,[1] MICHAEL W. RICHMOND,[31] CLAUDIO H. RIVETTA,[2] CONSTANCE M. ROCKOSI,[1] KURT RUTHMANSDORFER,[2] DALE SANDFORD,[6] DAVID J. SCHLEGEL,[4] DONALD P. SCHNEIDER,[32] MAKI SEKIGUCHI,[17] GARY SERGEY,[2] KAZUHIRO SHIMASAKU,[12] WALTER A. SIEGMUND,[3] STEPHEN SMEE,[5] J. ALLYN SMITH,[25] S. SNEDDEN,[7] R. STONE,[18] CHRIS STOUGHTON,[2] MICHAEL A. STRAUSS,[4] CHRISTOPHER STUBBS,[3] MARK SUBBARAO,[1] ALEXANDER S. SZALAY,[5] ISTVAN SZAPUDI,[33] GYULA P. SZOKOLY,[5] ANIRUDDA R. THAKAR,[5] CHRISTY TREMONTI,[5] DOUGLAS L. TUCKER,[2] ALAN UOMOTO,[5] DAN VANDEN BERK,[2] MICHAEL S. VOGELEY,[34] PATRICK WADDELL,[3] SHU-I WANG,[1] MASARU WATANABE,[35] DAVID H. WEINBERG,[36] BRIAN YANNY,[2] AND NAOKI YASUDA[16]
(THE SDSS COLLABORATION)
Received 2000 February 13; accepted 2000 June 8

FIGURE 24-4 Sloan digital sky survey notes.
SOURCE: Reproduced by permission of the AAS.

York, *et al.*, has about 3,000 citations, but has millions of downloads from the on-line database. There need to be ways to measure the impact of things like the SDSS and not just citations.

The next paper below is a secondary catalog of clusters created from the SDSS. This is the part we cited but the catalog itself is not in the journal. The catalog is quite large, more than 50,000 clusters, with about a hundred measures per cluster.

THE ASTROPHYSICAL JOURNAL SUPPLEMENT SERIES, 191:254–274, 2010 December doi:10.1088/0067-0049/191/2/254

A GMBCG GALAXY CLUSTER CATALOG OF 55,424 RICH CLUSTERS FROM SDSS DR7

JIANGANG HAO[1], TIMOTHY A. MCKAY[2,3], BENJAMIN P. KOESTER[4], ELI S. RYKOFF[5,6,12], EDUARDO ROZO[7,13], JAMES ANNIS[1], RISA H. WECHSLER[8,9], AUGUST EVRARD[2,3], SETH R. SIEGEL[2], MATTHEW BECKER[10], MICHAEL BUSHA[8], DAVID GERDES[2], DAVID E. JOHNSTON[1], AND ERIN SHELDON[11]
[1] Center for Particle Astrophysics, Fermi National Accelerator Laboratory, Batavia, IL 60510, USA
[2] Department of Physics, University of Michigan, Ann Arbor, MI 48109, USA
[3] Department of Astronomy, University of Michigan, Ann Arbor, MI 48109, USA
[4] Department of Astronomy and Astrophysics, The University of Chicago, Chicago, IL 60637, USA
[5] Physics Department, University of California at Santa Barbara, Santa Barbara, CA 93106, USA
[6] Physics Division, Lawrence Berkeley National Laboratory, Berkeley, CA 94720, USA
[7] Kavli Institute for Cosmological Physics, The University of Chicago, Chicago, IL 60637, USA
[8] Kavli Institute for Particle Astrophysics and Cosmology, Physics Department, Stanford University, Stanford, CA 94305, USA
[9] Particle and Particle Astrophysics Department, SLAC National Accelerator Laboratory, Menlo Park, CA 94025, USA
[10] Department of Physics, The University of Chicago, Chicago, IL 60637, USA
[11] Brookhaven National Laboratory, Upton, NY 11973, USA
Received 2010 August 5; accepted 2010 October 26; published 2010 November 23

FIGURE 25-5 Cluster Catalog entries form SDSS DR7.
SOURCE: Reproduced by permission of the AAS.

As the catalog is not in the journal, the question then is, Where is it? I did not know, so I queried Google and it showed me that it is on the personal website of the first author. This is it where we got it from. This is the type of data that is not linked to in the ADS.

So, these are two problems in how astronomers link their data.

References

(1) Astrophysics data system workshop. Workshop report, Annapolis, Maryland, August 18-20, 1987, Pasadena: California Institute of Technology (CALTECH,CIT), Jet Propulsion Laboratory (JPL), 1987, edited by Squibb, Gael F.

(2) Squibb, G. F.; Cheung, C. Y. NASA astrophysics data system (ADS) study ESO Conference Workshop Proceedings, No. 28, p. 489 - 496 http://adsabs.harvard.edu/abs/1988ESOC...28..489S

DISCUSSION BY WORKSHOP PARTICIPANTS

Moderated by Bonnie Carroll

DR. MINSTER: This comment is for Anita de Waard. You said correctly that commercial software producers do a good job. However, give me any piece of software and I will show you that some of its thousands of files dating back to 1995 cannot open and will never open again. On the other hand, my Linux mail from the 1980 is perfectly fine and I can open all my files with no problem.

DR. DE WAARD. You are right. I should not have said "commercial". I do not have any preference for any kind of software development. My point is not about how the software is being developed, but about the fact that software developers–commercial or academic—have an important role to play in the infrastructure of scientific communication.

PARTICIPANT: My question is for Bruce Wilson. The last time I looked at the DataONE project, they were using pieces of software from Mercury to obtain metadata from different sources. Is Mercury now producing outputs that can be used directly as a citation?

DR. WILSON: What I showed with COinS (Complex Objects in Spans; see http://ocoins.info/) is embedded in the Mercury results. It is the software that drives the search interface for the ORNL back and about 15 other data centers. It is also being used for search in the DataOne project. So yes, we are getting there in terms of using COinS in the search results. The package itself is also using the Open Archives Initiative-Protocol for Metadata Harvesting (OAI-PMH) and we have been extending it to expose OAI-PMH to other harvesters.

PARTICIPANT: Does it produce output that can be used directly as a citation?

DR.WILSON: COinS produces output that can be used directly as a citation in the sense that what we are trying to do is to provide structured metadata in a format that can be used by citation tools.

PARTICIPANT: My comment is for the commercial publishers. At least in my scientific community, biodiversity, there is a growing understanding that one of the ways to encourage the citation approach is to publish the data and to provide some incentives, something like a data paper. Given the fact that there would not be any operational burden on them, how do you think the commercial publishers will respond to such a call from the community? Would they produce a section in their journals dedicated to data papers where the datasets are described through the offering of metadata? Will the journals be engaged in the peer review and publishing of such data papers? It is important for scientists to publish and make data available in the open public domain, and therefore I think it is important that the commercial publishing community come forward and introduce such sections in their existing journals and publications. Do you think publishers would respond to that? How would this affect their business and operational models?

PARTICIPANT: I think the peer review of datasets is a very difficult task. If we look at Michael Kurtz' data, for example, we will realize that it would take a lot of work and time to do a good job reviewing the data. Someone needs to do that work and it costs money. This seems like the kind of work that governments usually fund. I do not think it is the publishers' job. If a journal agrees to do it because it is critical in their field, then that would be great. I do not think there is any blanket statement to be made here except that everybody realizes it is a difficult, time consuming, and expensive process and that somebody needs to pay for it.

PARTICIPANT: There is a good example from chemistry. There is a leading institute called the Beilstein Institute that does a lot of work in the data curation area.

PARTICIPANT: There is also a commercial version of that. I think that we need either a strong mandate backed with funding from a government organization or a private business model to make sure that the data curation is done professionally and properly.

PARTICIPANT: There is a journal that does that. It is called *Earth Science Data*. I think it has been marginally successful. Like any journal, they struggle to get reviewers but in this particular case, it has been more difficult and challenging.

PARTICIPANT: I should start by saying that I am a total cynic. Yes, we do have this session with a group of stakeholders in the research enterprise, and yes, data citation is the main theme, but I do not think that citation and attribution are going to solve all the important issues across this large spectrum. There needs to be a lot more outreach and interfacing. This is just an observation, not a criticism.

This question is for Bruce Wilson. If I understood correctly, you said that downloads of data did not necessarily correlate with the citations of data. If you look at the literature, however, downloaded papers do actually correlate with the citations of these papers to some degree. It seems to me that there is a difference of views here and I think it would be important to understand why that might be the case. Whether it is because the data are not properly attributed, because there is not enough metadata, or it is a function of different disciplines, it would seem to be an important point to understand.

DR. WILSON: I am not aware of broad studies on these issues. My statement was based on some observations of the roughly 1,000 datasets held by the ORNL. This sample has some limitations, but what we found, through simply going out and asking questions, is that there were cases in which people stopped working with the dataset because it was too hard or because there was a problem with the data. These findings helped us to identify some of these issues and fix them or greatly lower the barrier to the datasets. After fixing some of these issues, the number of downloads increased and early indications suggested that the citation of that data has also gone up. We have seen cases where some of these datasets are now being routinely downloaded and used in the classroom for undergraduate education. That is also another issue, where downloads might be attributed to other kinds of uses of the data. That is why I am interested in the discussion about what are the impacts of the data outside of the scholarly community literature.

DR. KURTZ: Downloads correlate with citations only when researchers use them. This is true because practitioners do not cite data. It seems that materials that are useful in practice are often never cited. Also, not everybody who downloads the dataset is planning to immediately write a paper. There are many other uses of data. If you look at download statistics in Google scholar versus citation statistics, there is no correlation, whatsoever, but if you look at download statistics for research articles by astrophysicists through ADS, the correlation is perfect. So, it really depends on what the use is.

PARTICIPANT: I have two points. First, I am following up on Phil Bourne's point about whether we are overloading data citation. I think that despite our attempts to keep this meeting within a narrow range, we also wanted to surface the whole set of issues concerning data citation and attribution. Second, I want to emphasize that describing the data for future reuses within the immediate discipline is hard enough. Describing it for future uses in adjacent disciplines and beyond requires much more context. Basically, the farther you want to go from the point of origin, the more interpretation is going to be required.

Potentially, this might be a librarian's full-time job. Allen Renear and I are among the few people in this room who built courses and educated libraries around data archiving, but I do not see several dozen of our graduates being hired for these jobs. I am not seeing the growth yet. There are real infrastructure and human resource issues here. If the panel could address whom you are hiring to do this job and why, that would be really helpful, too.

DR. CHAVAN: I want to comment on Michael Kurtz's point that citation is not the only way to measure the usage of the data and that there are several uses of the data that often do not result in scholarly publications. The way to address this issue in our community is through building what we call "a data usage index". This is an index with several parameters, whereby download and use of data is one of the aspects of the data usage index. Several co-authors and I proposed this index through a paper in 2009. The index has gone through community consultations and over the past 18 months and we will be advertising the algorithms auditors. We believe that this algorithm can be modified for different disciplines because of the different data usage patterns in different disciplines. So, in addition to data citation, we also need to promote and facilitate the creation of other forms of impact measurement.

DR. SMITH: Yesterday, someone talked about the importance of data citation to provide credit to researchers and that professional data centers also care about getting that credit. However, I do not hear that universities and libraries also require credit for the incredibly labor intensive work that they do to get the data managed and archived. It would be useful to discuss whether research universities and libraries also should get such credit.

PARTICIPANT: I think it is important to parse this topic well here. Let us start from a community perspective. In some communities, there is very clear value from having shared data and in this case there is an absolute requirement for standards for data attribution and citation. In other communities, on the other hand, the cost-benefit analysis does not come out so clearly in favor of benefits.

The second point I would like to make is about credit. Some institutions choose to play a role in the community in the provision of data assets that extend beyond the institution. In those cases, it is very clear that those institutions expect to be rewarded for the role they are playing. When it comes to getting credit for having, for example, an institutional repository that serves the need of the researchers in one institution, I would not expect that institution to require credit for that work. This is simply the responsibility of this institution to its stakeholders. That maybe helps to tease out some of the issues in the discussions we have been having over the last couple of days, because when it comes to data attribution and citation, I do not think there is a one-size-fits-all model.

DR. WITT: From a library perspective, I think it is more about supporting an institutional mission more so than credit. It is also about relevance. Books and journals are going away and there is more focus now on how libraries are going to deal with data collections. So if we are not an actor in this process, whether through discussing data management plans, building repositories, or creating services to help people find and use data, libraries will lose relevance. As for the workplace and the kind of new organization or infrastructure that the libraries will need, I think that the principles are all there but the technology and the packages that are in place need to evolve to meet the requirements of the new tasks.

Take cataloging, for example. The people who understand the AACR2 (Anglo-American Cataloguing Rules) and how to create records and descriptions will still very much be needed. Such areas are still relevant to the data world. It is not what needs to be done, but how it is done to make these changes. We can look at the organizational chart of libraries and try to identify the relevant components, whether they are in informational literacy outreach, collection development, metadata and technical services, acquisitions, or archives. I currently see many libraries defining new positions related to data service and curation. So maybe we should take a hybrid approach, where a library will have folks who do the new data related work, and other folks, who still work on more traditional areas.

PARTICIPANT: At our publishing house, we are hiring knowledge modelers. This category can include different domain specialists, such as economists and technologists. They do modeling, analysis, and visualization.

MR. UHLIR: I just want to point out that the NRC Board on Research Data and Information will be doing a consensus study on the future workforce and educational requirements for digital curation, starting in the fall of 2011. We will be looking at all of these issues.

PARTICIPANT: I think that what will happen in the data citation context is similar in some ways to what happened when we went online and searching online became an important skill. Libraries needed the expertise, so they hired some specialists who did searching for their patrons. In the library schools, they hired experts who did specialized courses on online searching that were very popular.

DR. WILSON: The one thing that I would add from a data center perspective regarding workforce needs is that we are frequently looking for what Mark Parsons has called the data wrangler, which is somebody who has domain expertise, information science expertise, and understands that what they are going to be doing in ten years, even though it may have nothing to do with what they are doing now.

26- Standards and Data Citations

Todd Carpenter[1]
National Information Standards Organization

Standards are very familiar in the distribution of information, even if we do not recognize them. They are things that we rely on every day. We probably do not give them much thought. However, when we pick up a book, we are really picking up an incredibly standardized object. Everything from page numbers, table of contents, an index, cataloging information, title pages, organization structures, paper acidity, binding, paper sizes, ink, colors, font sizes, even the spelling of the words themselves are all best practices derived from the mass production of book publishing. These practices have developed over time and have been adapted to provide more efficient discovery and distribution of content.

One challenge posed by our current environment and the transformation to digital content distribution is that a lot of the practices that have been resolved for decades have changed radically in this new digital environment. If we think about page numbers, for example, they do not really mean anything in the flow of a digital text, where you can size the font up to 96 points and a page might only have two or three words on it. So, how one cites those particular words within such a mutable digital object is certainly a problem.

As I envision the citation of the future, it is no longer a string of words and textual descriptions of how to discover a referenced item. We are moving toward an era when a citation will likely be one or more identifiers for the specific information being referenced.

The information distribution ecosystem is moving toward standardizing around a number of key identifiers. These identifiers provide us with actionable answers to a variety of questions: What is this thing? What is its relationship to other things? Who created it? What is it packaged with? How can I locate it? If we can create a citation structure that is built on actionable links to entities that have additional information stored in accessible registries, there is great potential to aid discovery and distribution of information. After all, an actionable network of linkable text citations was, in part, the rationale for creating the entire World Wide Web and its follow-on the Semantic Web.

One of the barriers to establishing this network infrastructure is that the machines needed to intermediate this world for us do not talk the way that human beings do. They rely on structure and markup and are not able to easily gloss over errors in coding, syntax, or semantics. However, they are incredibly well suited to navigating a structured world with incredible efficiency. This fact has implications on the access, use, and citation of data. Instead of using a text-formatted description of an information resource, an identifier-based citation format, built on universally adopted standards, could build upon and unleash the opportunities in our machine-intermediated world.

---

[1] Presentation slides are available at http://sites.nationalacademies.org/PGA/brdi/PGA_064019.

We are almost to a point where we have identifiers for all of the elements of a citation. Identifier standards exist, or are being developed, for names (ORCID or ISNI), affiliations (Institutional Identification, I^2), publications (ISBN or ISSN), collections (ISCI), and persistent URLs (DOI, ARC, or PURL), and dates. Each of these standards could be incorporated into actionable URIs and those URIs, along with the associated metadata, could be served to the community as part of linked data stores. The implications of this shift of meaningful connections and machine references to a wealth of additional information could be tremendous. For example, an unambiguous name identifier could bring a user more than just the name of the referenced the author; it could also provide links to everything else this author has published. Another link in that URI-based citation could connect to everything else in this package or collection.

On several occasions during this meeting, we discussed the development of the Open Researcher and Contributor Identification (ORCID) initiative whose goal is to establish a unique identifier for each researcher in the scholarly communication process. This project is closely related to the International Standard Name Identifier (ISNI) standard (ISO 27729). This standard was recently approved for publication and it defines an identifier for any "parties" involved in the content creation process across all media. Both of these initiatives will probably launch in 2012 and will provide us a great opportunity for uniquely identifying content contributors and clearly distinguishing between people with the same or similar names. NISO's own Institutional Identifier (I^2) project will be utilizing the ISNI and its infrastructure to identify institutions and to provide metadata about them, including its links to parent or sub-organizations, such as departments. Combining these new identifiers with existing standards, such as the ISBN or ISSN, we are approaching a time where all of the information in a citation can be replaced with URIs.

So, when we talk about standards for data citation, what do we mean? There are a variety of things we could standardize that are related to, for example, discovering the data locating it, describing it, sharing it, preserving it, and for interoperating with it. But which are the most important to pursue? The problem with setting priorities is that each person or each field has different challenges and needs. What is a critical issue for one community is of secondary or tertiary concern to another. Here is my list of the things that I believe is being a high priority for good citations in a digital world:

1- Disambiguation of the item.
2- Location of the item (either in physical or digital form or both).
3- Attribution and disambiguation of the author.
4- Ability to reuse and preserve.

You may think there are other priorities; for example, ontologies and terminologies, privacy issues, rights and intellectual property issues, database size and complexity, and refresh pace and update frequency.

Identifying the most critical needs is really the first step. Mark Parsons said it well yesterday: if we can solve 80 percent of our problems with an 80/20 solution, we should do that. In large part, that is what standards do. Perhaps the data citation group that organized this meeting should spend some time focusing on which issues are secondary to the bigger goal of sharing data and then focus its attention on those things most critical to creating a culture of digital data citation.

Focusing on the core problem does provide a great deal of benefit even if it does not solve every issue for every community. One good example of this is the Dublin Core metadata standard, which is widely used. Communities have been trying to develop a better schema for how to describe content, metadata, and bibliographic information and they keep going back to the Dublin Core and extending its basic model. The Dublin Core metadata set contains all the critical elements and that same approach is probably how we should start. Perhaps what we need is a good framework of data elements that define core and secondary sets of metadata for describing data or data sets, which may be applicable in across a range of communities and leave the proper display question to each individual community.

The problem with most standards is not that they are bad ideas or that there is not sufficient thought that went into their development. Frequently, the problem stems from a limited amount of follow-up or commitment to promoting adoption. The point at which most standards fail is after consensus is reached. The failure point is often in the standard's adoption—or better said in the absence of adoption.

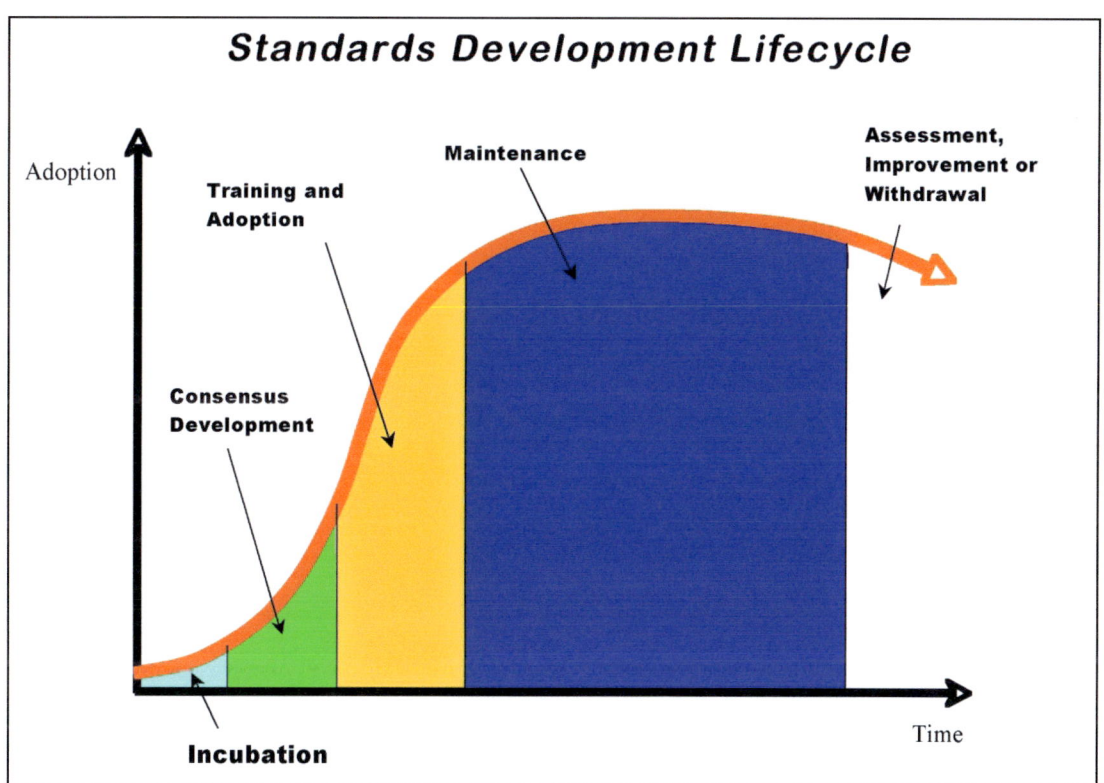

FIGURE 26-1 The standards development lifecycle.

Much like the sales adoption cycle of any other product, standards go through a lifecycle process of slow growth, growing adoption, broad acceptance, then eventual decline, which leads to revision or withdrawal. Figure 26-1 is a graph of the life cycle of a standard with adoption on the vertical axis and time on the horizontal axis. Projects start out in incubation stage, where a small group of people are interested in trying to solve a problem. Then there is a period of consensus development, where people become aware of the project and agree on the solution.

Unfortunately, far too many standards development processes stop at this stage. Group members might say or think, "We have developed the schema, so let us publish it, and then we are done." But the problem is that getting people to use a standard is the most difficult part of the process. Just as with the latest technology product, in order for it to reach mass-market appeal, it needs promotion, marketing, and encouragement to get people to use it. So, we need to spend a lot of time, effort, and energy on these later stages to make a standard succeed. If we do come up with a standard data citation format, schema for metadata about data sets, or publication policies, and after we have obtained consensus on the citation structure, we have to invest time and effort in getting people to actually use it.

Finally, I want to reflect on who is the audience for the project we are developing? "Who" is one of the most critical questions regarding the adoption of a standard or best practice that has been developed. Among those who need to be deeply engaged in data citation standards adoption are researchers, educators, data centers, publishers, promotion tenure committees, administrators, funding agencies, consumers of the data, and repositories. Among the challenges as we move forward is ensuring that we have engaged the right communities to ensure that what we have done here and what we will continue to do over the next year will get adopted? I think that is probably the biggest challenge facing us now, as it is with any standards development organization, or any standards community.

27- Data Citation and Attribution: A Funder's Perspective

Sylvia Spengler[1]
National Science Foundation

I should start by saying that I do not speak for all of the funding agencies and to emphasize that there may be differences of opinion within the National Science Foundation (NSF) itself regarding the issues being discussed here.

NSF cares about data citation and attribution for a number of reasons. A primary reason is that the United States Congress pays special attention to what happens in science and wants to see value for the money it allocates for science and education. That is a major determinant of why we want to encourage people to provide citations for their data—because it makes this effort more visible. It also helps convince the taxpayer, the people who actually provide the necessary funding, that there are good things coming out of this investment. I also believe that making data citations clear and a common practice will help promote the cutting edge interdisciplinary research, which in turn will help people in their career development and make their contributions to science and to the public good more visible and appreciated. The fact that the National Science Board is actually engaged in the issues of data policy, data citation, and data access gives us a big incentive as well.

Let me now talk briefly about what we are doing at NSF. Everybody knows about the requirement for having a data management plan in the proposals submitted to NSF. It is important to note that we recognize that one size does not fit all. That is why individual review panels and their managers make decisions and recommend proposals for funding on a case-by-case basis. Let me give an example. I had a panel in which everyone liked the intellectual merit of the project. Everyone thought it had incredible broader impact in terms of education. The principal investigator (PI) had cited his/her web page data policy. The panelists went to look at the webpage for data policy and said, "We think this is an intellectually stimulating and engaging idea that has incredible education outreach but because of their data policy, we do not recommend funding it." The PI was very responsive to this evaluation and I am sure that this will happen more in the future.

We are also introducing some changes to the annual and final reports to recognize data contributions, specifically to recognize individuals' role in data maintenance. Finally, one of the pieces that PIs have to provide when they write a proposal to the NSF is what they did with the money we gave them the last time. They must have the results of their data management plan (i.e., data access, preservation, use, and so on) available and listed in the references to stand higher chances of getting more funding.

Everything I will say now about the data management plan gets highly specific, sometimes at the program level, at the provision level, and at the directorate level. Also, solicitations may have additional data management requirements. The NSF Policy Office has a searchable website that

---

[1] Presentation slides are available at http://sites.nationalacademies.org/PGA/brdi/PGA_064019.

links to relevant guidance documents and examples. It is available at: http://www.acpt.nsf.gov/bfa/dias/policy/dmp.jsp.

The America Competes Authorization Act that passed at the end of December 2010 required the formation of federal interagency groups to discuss two major issues: public access to publications and the data supported (in whole or in part) by federal funds. A group on digital data at the White House Office of Science and Technology Policy is specifically looking at data policies and data standards. I also want to underscore the role of university and other institutional libraries and repositories, not only in acting as repositories but in actively developing systems for dealing with what everyone recognizes as a major challenge of metadata, including minimum metadata, usage generated metadata systems, software metadata, and the like.

I want to acknowledge as well schools of information science, which are helping to develop protocol software and systems that we use. The scientific societies also need to be acknowledged, since they are becoming clearer in their ethics statements and in their expectations for membership about the necessity of having not only citable publications, but also citable data.

Let me conclude by summarizing what I have heard over the last two days:

- Basically, citation is a fundamental ethic in science and it is the right thing to do.
- There is a great enthusiasm and support for data access, sharing, use, and citation and attribution.
- Technologies, per se, are not an urgent problem. It is the cultural and sociological challenges, since one size does not fit all and nobody pays attention to the instructions.
- We also should remember that there are both human and non-human communication mechanisms that need to be taken into consideration.
- We should not wait for the perfect solution for the issues under discussion: individual communities are making some good progress and they should collaborate and coordinate.

Finally, I would like to emphasize that I am interested in the different ways in dealing with granularity across different communities. I think this is an important issue about which I would like to hear some more discussion.

DISCUSSSION BY WORKSHOP PARTICIPANTS

Moderated by Christine Borgman

PARTICIPANT: I want to ask about the bottom-up standards approach, best practices, or conventions. I have heard a lot over the past couple of days about what seems like a growing convention on how to do data citation. What we have seen in some of our work is that whenever there is a convention that emerges, what we often do is invent the standards and then have to redo them so that we can embrace the convention that has been adopted. Maybe someone could say something about what you think about data citation and convention.

MR. CARPENTER: One of the issues with standards development is that if you are too forward thinking, people will not get behind it. Sometimes it is better to let an ad hoc specification begin in a particular community and after it has gained some traction, move it into formal standards development for a broader audience. Such an approach can be very useful because, ultimately, it is all about adoption. Standards will not be helpful if they are not being used. Part of the process should be getting the community's buy-in. I know it is a big problem, but it is a matter of timing and marketing.

We have found with different standards that often what makes a standard popular is an application that shows the different things that you can do with it. I do not know what the best demonstration application might be for data citations and would like to know if someone has ideas in this regard.

DR. SPENGLER: One of the things that I have noticed is that when major leaders in the scientific community, whether it is research funders or journal publishers, have some requirements, it often helps with standards. So, if this group, for example, comes up with some recommended standards for the data citation, it might be useful to see whether or not some organization like the National Science Foundation (NSF) would welcome that. This might be one way to make the transition.

PARTICIPANT: If someone were to write a proposal based on the discussions today and send it to the NSF, to what program should it be submitted?

DR. SPENGLER: I do not represent the entire NSF, but I would say either Mimi McClure from the Office of Cyberinfrastructure, or me from the Directorate for Computer and Information Science and Engineering. It would fall between the two of us.

PARTICIPANT: I want to make a suggestion related to standards and the usefulness of data citation. It would be good to be able to check the dataset and make sure that it was not changed since it was downloaded the first time. This would allow us to know if the generators of the dataset found anything wrong with it and if that they have recalibrated it.

PARTICIPANT: I will ask a policy question. The NSF's approach with the data management plan is to enforce it via the proposed review process on the front end and then the reporting requirement on the back end. The National Institutes of Health (NIH) has had such a data management plan requirement for large grants over a half million dollars and the plans have not been part of the peer review process, just between the investigator and the program officer. The Economic and Social Research Council (ESRC) in the UK has gone a very large step further and requested that to submit any proposal to gather new data, an investigator must show that no other data exist that he or she can already use. This is a whole different kind of policy. What would happen if we tried to do something like that in the United States? That would certainly be a game changer.

DR. SPENGLER: Yes, it would be a game changer. The question is how would you certify any of what the UK is requesting? Is this accessible from my university? Is this accessible with the adequate permissions? How can it be accessed? I think that the reason for the NSF to go for the review process and to include the community is because communities are part of NSF's highly individualized approach to funding science. Program directors at NSF, except for some, come and go based on the two-year and three-year rotation model. What we want to do is to engage the community. We do not want to make it a top-down approach. We want to make it bottom- up because that is our tradition and we want to have communities make clear what is adequate for them. I could possibly take the standards that the genomics community has for data and use the same approach for people who necessarily spend large portions of their lives in less than amiable environments, trying to push forward other areas of science. That would not be very fair of me as a program director or as a reviewer. I have to think about what my rights are versus their rights.

PARTICIPANT: The ESRC requirement to look for previous data was interesting. At one point, and I do not know if it is still the case now, the Department of Defense required that in order to do additional research, researchers had to prove that they searched the literature. Most people do read the literature and that is why they have bibliographies when they are embarking on new research. They have to prove that they have searched the literature and there are systems to do that effectively. Until we have good data repositories (i.e., clearinghouses, so we know how to find what data exists), it is going to be hard to request the same thing for data. It is the data discovery tool that we do not have yet.

PARTICIPANT: I mentioned yesterday a catalogue of many resources in the bioscience area. We obtained all the URLs and their papers in each issue. The attrition rate was about 10 percent per year. There seems to be some conflict between requesting researchers to deposit data and making more data available while they do not have the repositories they need to actually carry forth the policy. How is the NSF addressing that situation?

DR. SPENGLER: The Directorate for Biological Sciences has put its resources on infrastructure in a variety of different places, but there is not any activity that is funded to do that specifically. It is a leadership challenge within the different directorates. Availability and preservation are two very different issues and it is not at all clear to me how that is adequately dealt with and that is why I am speaking as Sylvia Spengler, not on behalf of the NSF.

MR. CARPENTER: I think that we as a community are not investing enough time, effort, and particularly money in long-term preservation of content in all forms, not just data. For example, if a library holds a book, you can expect that that library will keep that book until eventually they run out of space and even then, you might still be able to get the book from some form of repository. We do not do that with electronic information. We are increasingly in an environment where we lease content from organizations, but we do not own it. I think we might get to a point where we are living in a digital black hole a hundred years from now because we are not investing time and resources in preservation.

PARTICIPANT: One thing that we have seen from private foundations in recent years with regard to the sharing of physical materials is to require researchers to demonstrate that the research that they are doing is novel. They will only give funding and access to some physical material resources, such as blood samples or spinal samples, if the researchers demonstrate that it is truly novel research, not just incremental. Then researchers have to share the data back. This is something that we are starting to see some private foundations do.

PARTICIPANT: One thing the big funders might need to consider is to create a condition in which universities and research institutions accept inbound policies from smaller funders, because there are 2500 disease foundations in the United States alone but very few of them can fight Harvard to mandate a data sharing plan, format, or standard. Guidance to those foundations and non-traditional funders can be very powerful in facilitating adoption in this difficult period where well-funded scientists at top universities are not going to take that money, but a scientist at East Tennessee State might look for such funding and adopt the standards as part of the deal. Having the big foundations and funders lay the groundwork for adoption of that broader policy would be very useful.

MR. CARPENTER: That is a good point. I think there are a variety of communities engaging in a very traditional landscape. Keeping in mind who those new players are and how they communicate would be very useful.

PARTICIPANT: This question is for Sylvia Spengler. I know that the NSF requirement for the data management plan is new, but I am wondering if there is any experience regarding reviewers and panelists, how they are accepting this added responsibility for reviewing the data management plan, and whether they feel they have adequate training to do it?

DR. SPENGLER: We actually have developed sets of materials to address these issues, both as instructions to reviewers when they start looking at the proposals and during the panels themselves. There are many issues involved here. There is an education process within the NSF for the program directors so that they become aware of the importance of these data plans. I think that part of the reason why it took so long to make the data management plan requirement visible is that there was a lot of concern about the additional effort that it would require not only in review, but also in award oversight. My guess is that in the long run, that will turn out to be part of submitting an annual report and, as we all know, the annual reports and the final reports enable researchers to continue to receive money. The funding agencies are not opposed to being a stick when pushed to do that.

DR. CALLAGHAN: I thought I would give a different example of what is happening across the funding agencies in the UK. Most of the money that is funding my work today comes from the National Environment Research Council (NERC) and they are very keen on implementing data citation and publication. They also released the new data policy in January 2011, which essentially states that all data collected under research funded by them should be made publicly available through publication and environmental data centers. That is a good thing as far as we in the data centers are concerned, but we still have to convince the researchers who produce the data to deposit them appropriately. The other Research Councils of the United Kingdom are following suit as well. There is pressure coming from the UK government, which decided a few years ago that if any scientific data or any data is collected as a result of public funding, it should be available to the public. So, there is pressure to do this, but it is up to us to tell NERC and the UK government what is the best way for us to get the data producers to comply with collecting, and then publicly sharing and archiving the data.

DR. BOURNE: When someone mentioned the "stick", it made me think of the NIH open access policies as something that could be considered for the NSF data requirements policy. It might be worth looking at how the NIH policy is working and what additional lines and budgets to support it are expected.

DR. SPENGLER: I must clarify something about the NSF access policy. At the moment, you can get to an abstract, but you may or may not be able to get to the entire article. It is clearly something that is on the table, however. That is why there is an inter-agency task force or working group at the Office of Science and Technology Policy trying to deal with questions of public access to both publications and data.

MR. CARPENTER: The publishing community is certainly interested in partnering with the data repository and scientific community because they recognize that they do not want to be performing those functions. The publishers are not interested in being the repository for any public domain data. It does not fit well with their business models.

PARTICIPANT: I think that the publishers are listening and they want an access policy proposal that cuts across domains, obviously. They will have greater difficulties with different standards for different domains. While they will understand a diverse situation, the more generic the guidelines are, the easier it will be.

MR. CARPENTER: As I mentioned earlier, there is a project currently within NISO to look at how to tie together whatever supplemental materials are submitted with a paper, be that a dataset, video, audio file, and so on. The publishing community is already thinking about this and trying to address some of these concerns and issues.

PARTICIPANT: I think the two key issues here are quality and discoverability. That is what the scientists and publishers care about.

DR. KURTZ: Besides quality, reusability is very important to the operation of standards. In the astronomical virtual observatory movement, what we call the International Virtual Observatory Alliance is basically a standards organization that is developing complex standards for characteristics such as at what time was the observation taken, what wavelengths are involved, and the like. It is a description of the observation so that it is machine readable and reusable by some kind of standard software tool. The increasingly complex data standards are clearly field-dependent, but they are necessary for machines to communicate and evaluate data so that people do not.

MR. CARPENTER: I think there is a difference between the very domain specific intra-operability question and the more general 80 percent answer to how do we find, locate, interact with, and discover data. As a community, we need to be careful not to tread too closely into the domain-specific area because it very quickly gets bogged down and we will not be able to accomplish anything if we focus too much attention on those 20 percent solutions that are very domain specific.

DR. SPENGLER: I would like to go back to the comment on quality and discoverability. The National Science Board has had discussions about using data citations for biosketches and resumés of Principal Investigators. One of the points that Todd Carpenter made was about peer review and I was pleased to hear this point brought up yesterday. However, the reality is that there is nothing in any of the citation styles that I saw discussed yesterday that says whether or not something was actually peer reviewed. Some researchers post their dataset online with very low quality. I know this is their issue, but where does the peer review come into the picture? I am hoping that the report that comes out of this workshop actually addresses that aspect.

MR. CARPENTER: One of the really interesting conversations that the publishing community has been having within the joint NISO-NFAIS project on journal article supplemental materials is the difference between what is "core to understanding" and what is "supplemental". If it is core to understanding then it should go through the same rigorous review process that the paper goes through. If the information is not really critical to understanding or is just supplemental, then the question is do we really have to review it—or even have it? This has actually been one of the most interesting philosophical conversations taking place among the publishers in the NISO project in terms of defining what is supplemental.

PARTICIPANT: I am glad you brought up the peer review issue again. There is nothing in the current citation practice and literature that implies peer review. It is all about norms. Depending on the discipline, different materials get different levels of review and it is all very norms-based. It is the sort of thing you learn through your career as a scholar.

MR. CARPENTER: In a print environment, we are relying on the reputation of publications such as *Nature* and *Science*, which has developed over decades. It is not perfect but we have a culture that has built up over time and we cannot simply replicate that today in a new environment because we have shifted to focus on data as opposed to publications. That is going to take additional time.

PARTICIPANT: We should separate concerns and try to solve some fundamental problems first. Citation and peer review are connected, but different. We have already heard that the journals that have started to do peer review of data are struggling. I want to point out that one of the current bases for the ranking of journals is how many citations refer to them. In the same way, we could start to build up a ranking system of the data centers, if that is a necessary outcome. The first step would be being able to count and track the number of referrals to a data center. I think that probably could be solved by concentrating on the citation element and then the quality of particular data centers would come out through those numbers and through other practices that are yet to be defined. There is a way of approaching this in small steps.

PARTICIPANT: I want to comment on the point regarding over-reliance on the notion of peer review. When we have some of the larger fields with shared instruments like astronomy, that is very different from the folks who are in small areas of ecology. We do not have the kinds of agreed upon databases in all fields. Those of us who like to call themselves inter-disciplinary sometimes publish in computer science, social science, and information studies, for example. I publish both quantitative and qualitative work in these fields. I cannot even tell you who the peers are who would examine my data. There is consequently a huge long tail of fields where the community is not clear to develop its standards and policies. I am concerned that we are using peer review and community in a sense of big science, rather than this long tail.

MR. CARPENTER: The peer review process is community-based and the review criteria for computer science, astrophysics, and biology, for example, are somewhat different. If we have a database in a particular field that is core to our understanding, then it should go through the same process that a paper in that field goes through.

DR. DE WAARD: I am wondering why the concept of "core to our understanding" seems totally wrong to me. It seems that there might be different use cases of data and it might be good to differentiate among them. One case is when you are convinced that the story that the author is telling is true. You need to look at the data and how they were obtained to be convinced. In this case, we can say that data are core to our understanding of the paper. There are other use cases and strong arguments for depositing data, however, even when it seems perhaps trivial for the authors themselves. This might allow others to do other types of research if the data are deposited in a usable format. Gully Burns proposed to deposit data in such a way so that someone can actually have meta-studies that cut across different types of research. Another example is Einstein, who looked at Michael Morley's work because he was able to access the data that they could not interpret and this offered support for the theory of relativity. I think it is important to recognize that there are different use cases of any datasets.

DR. BOURNE: I want to reiterate that talking about data citation together with peer review seems a very big activity and maybe something that should be addressed separately. If you look at the peer review of papers, the strain on that process is unbelievable. I get many requests to do peer review and I do not think I could do it for data. I can determine whether the data are good or not only when I use them.

DR. SPENGLER: People who get data online frequently have an almost instantaneous reflex to find out who funded the data and report any usability or quality issues. Whether or not we consider that as act of peer review is open for discussion, but it does happen and you would be surprised how long people remember that they could not use a dataset.

DR. CALLAGHAN: When it comes to peer review of data, we have been thinking about different levels of citations. We have what we call "plastic citation", which is the case of researchers simply putting their datasets on Excel spreadsheets and posting them on their personal web pages. It might not be usable as far as other users are concerned, because they might not be able to open the spreadsheets, but the datasets are citable. The next level that we call "silver citation" is when the dataset is in a repository that is generally trusted by the members of the community. Here, we can make certain assumptions about the quality of that dataset simply because it is hosted in a repository. If we have done our jobs properly, the mere fact that it is there and cited means that it is in an appropriate format. Even if the format is going to be migrated or changed, the metadata will be there and will be as complete as we can make it. Moreover, when you open the file, you will be able to do that using standard tools. So, by the mere fact of the data being in a trusted repository, we are more confident about them. In terms of technical aspects, this is actually going to be quite helpful for the scientific reviewers because they know that if it is in the right repository, they would not have to worry about finding the right program to open the files.

As for the scientific peer review itself, given that technical issues are taken care of, reviewers can focus on the quality, value, and other important attributes of the dataset. So, in a sense, we have got two levels of peer review. We have got the technical peer review, which is done by the data centers, and then we have got the scientific peer review, which is done by the domain experts as part of what we consider the formal scientific journal publication process.

PART SIX
SUMMARY OF BREAKOUT SESSIONS

Breakout Session on Technical Issues

Moderator: Martie van Deventer
Rapporteur: Franciel Linares

The breakout session on technical issues for data citation focused on synthesizing and bringing together ideas from the individual participants. The purpose of this breakout session, like that of the other three breakout sessions, was to:

- Identify the key issues that were raised during the workshop.
- Identify those issues that are important to the topic that were not already discussed.
- Discuss in greater depth the issues that the breakout group thinks are most important.
- Identify several issues for further work and choose one for discussion in the plenary session at the end of the workshop.

What are major technical issues related to data citation, what are those that were not discussed and which ones may be more important? The group created a list of issues that were considered important regardless of the order in which they were discussed:

- Determining the right versions to cite;
- How we can use existing web conventions, such as landing pages;
- The need for standards in creating human-readable and machine-actionable landing pages (e.g., XML, RDF);
- How this relates to web mechanisms and how to leverage existing paradigms and not reinvent the wheel;
- The need for a of a set of examples of existing technologies that are being used to illustrate;
- How to put a dataset into a bibliographic tool, including syntax;
- Views of data citations, and how to identify referenced datasets, granularity, and subsets;
- How to determine what is really being identified (e.g., the item itself, a descriptive landing page, or a journal article or the XML document representing that article);
- Identity (including scientific equivalence);
- Granularity of the database and citation; and
- Location versus identification.

A data citation might be something that simply identifies the data that has been used, or it might also provide a means to access the data. One key issue identified by several participants is how a data citation included in a paper might deal with both the identification of the data used (which has implications both for the issue of credit and for the issue of scientific reproducibility) and also provide a means to locate the data—which is often crucial to provide assurance of scientific reproducibility.

The group discussed the necessary and sufficient characteristics of an identifier. From the DataONE research program perspective, the only necessary characteristic is that it be unique (within a particular name space). An underlying issue is the level of the information that is being identified. MOD12QA1 collection 5 is an identifier for a particular concept, but does not identify the specific files used for a particular scientific analysis. It is an identifier that is sufficient for assigning credit, but it does not provide enough information to identify the source of the particular data. Several participants noted that an identifier in a citation should provide enough information, perhaps through use of a resolver service, to be able to get to an online "landing page" that identifies the data and perhaps also provide information about access of the data. At minimum, as one participant observed, an identifier should be unique, not reusable, and not transferable.

As several participants in the breakout observed, the data citation can have a number of characteristics. It can be in the format prescribed by the style guide used by the journal. It can include the DOI to (the intermediary "landing page" of) the data being cited, so that data centers can use it as a fixed string to search on to find uses of their data. It may also include a DOI to an author-created landing page, which for clarity we can call the "citation page", and which contains information about how the dataset was further divided, processed, or otherwise manipulated. It ought not link directly to the data. It may be in human-readable text until such time as we have domain standards to explain sub-setting, processing, provenance, and the like in machine-readable form. It can be stored for the long-term and have a DOI assigned to it. It also can link back to a data center's "landing page" for the data.

What is being identified could be clearly defined (e.g., a landing page or a journal article or the XML). In a data citation there could be an identifier that identifies a landing page about the dataset of interest and it could involve a resolution mechanism. Given that citation and given the associated identifier a user ought to be able to find the landing page that contains information about the dataset. Since the citation itself should not go to the data, a suggested access paradigm is that there could be something in between the citation and the data. The string selected for an identifier could be provided by an organization with long-term longevity, and which has the authority to do so.

Versioning is also an important topic. A Uniform Resource Identifier (URI) can lead a user to a landing page and then another URI that leads users to the granular version of the data. A certain version and granule may have its own URI. This information could be part of the citation created by the person who used the data. It reflects an attempt to recognize the scientific concept of the data that were used and also an attempt to encapsulate the specific subset of the dataset that was used, particularly for situations in which the entire dataset may not be completely reproducible.

One of the participants raised three cases that could be considered. In the first case, the citation would be simple and would always point to the original landing page (or something similar). This would establish credit, but would not refer to the specific instantiation used for the particular paper. (That is, it would not address scientific reproducibility sufficiently).

In the second case, each user could create (in some location with long-term longevity) a landing page that refers to the specific instantiation (granules, manipulations, subset) used in the particular study. This would handle the scientific reproducibility within the limits of the longevity of the landing page, but would complicate the issue of credit, depending on the infrastructure by which the dependency tree is created.

In the third case, one could have a more complicated citation, which would pull together both the original URI for the data and a URI that describes the specific instantiation used for the particular scientific study. This citation would be more complex, but would provide two URIs: one URI that could be used from a credit perspective and refers to the concept of the scientific sense of the dataset, and another URI that could point to the specific instantiation of the scientific data used for the particular study.

Another participant noted that the group supplying or publishing the data could provide a page of information about the data. There could be standards created for these pages that specify the minimum information necessary for basic citation use, but that would be extensible for domain-specific information, or other value-added services, such as linking to papers that use the data. Citations could go to this intermediary "landing page," rather than directly to the data, as the data may be excessive in size, and not useful without the proper documentation or software to read it. The page could have information about the data suitable to create a citation in the various different citation standards. The URL to this page could serve both human-readable and machine-readable information. There also could be a standard for the machine-readable portion, and it would be best to avoid competing standards.

The landing page could have information on how to obtain the data and how to use the data, such as links to papers about the data or the instrument that created the data, other grey literature documentation, or software necessary to read the data. The landing page could be stored for the long-term and have a DOI assigned to it, although the landing page may change over time. It could be updated when the data are moved, or are no longer available. If the data are replaced by a new version, a user ought to be able to link to a page describing the dataset as a whole (without versions), that would then link to the most recent version, rather than linking to the next version (so we do not have as long a chain to resolve to find the data). This page could be an appropriate place to give credit to all of the people who are involved in the creation, validation, and maintenance of the data.

There was further discussion of how to cite subsets of a dataset. One case could involve using the original identifier plus another identifier to describe the new subset. Do we complicate the citation in order to make giving credit easier or do we make it simple to create the citation? In the latter instance, pointing to the new landing page identifying the subset could make giving credit more difficult if there were no way to trace back to the original dataset.

If there is a landing page to a subset of the data, it could link back to the original data center, publisher, or landing page also. If it has been replaced by another version it could link to a landing page for the un-versioned dataset.

Questions that could be discussed more in the future

While most of the discussion in this breakout group covered data citations and how to identify referenced datasets, granularity, and subsets; other questions were identified by some participants as potential subjects of a subsequent meeting:

How should we handle the aggregation of datasets (e.g., data from over 100 sources)?

Some of the current bibliographic systems might search extended methods supplements for references. Is it sufficient to have only the data supplier's landing page DOI in the citation?

Would guidance similar to when you use "et al." be useful? For example, if one is citing three or fewer datasets, each one could be cited individually, but if four or more datasets are aggregated, then could one use the citation page to aggregate them?

If as part of the research, a new dataset is derived or synthesized, and is going to be made available, are the researchers obligated to cite back to the original data source, or just to their "new" data, which then might have the link back to the landing page from the data source?

Breakout Session on Scientific Issues

Moderator: Sarah Callaghan
Rapporteur: Matthew Mayernik

What are major scientific issues related to data citation, which ones are general and which are field or context specific? Before looking at some potential answers to those questions, some of the participants in the breakout group first tried to get a better grip on what a "scientific issue" meant. One thought was that a science issue is something that represents a disciplinary matter, in contrast to technical issues, which focus on how to do data citations. For example, determining what a "data aggregation" is that can be cited would be a scientific matter. This could be decided based on disciplinary community norms.

One participant noted that an important scientific issue is dealing with equivalence. The scientific equivalence of datasets is an outstanding problem because, ideally, the users of data would like to know when looking at two citations whether the citations are "the same thing" from a scientific point of view. It can be challenging because data often lead to derived data, or they may be subsets of larger citable data collections. This points to a couple of key scientific issues with regard to data citation, data versions (how to cite data that change), provenance (how to track that citations are to data that have not changed), and data linking (how to link to data that are poorly bounded objects).

Another participant observed that many scientists would rather be using their data than managing them. This does vary by discipline. For example, bioinformatics is strongly based on using data created by someone else, which implies that people are making their data available to others. A distinction might be made between disciplines that are based on using others' data versus discipline based on providing others with data. Different domains have different cultures, different funding mandates and norms, and different shared histories of practices. Some data practices are also very dependent on individual personalities.

Another issue deemed important by some of the participants is that scientists do not have enough time to do all of the data work that is necessary to make their data usable by others. At data centers, there are people specifically responsible for cleaning and archiving data. What kind of partnerships might be available between scientists and data archives or centers? At a data center, it can be very difficult to work with scientists, who may be reluctant to collect appropriate data or provide full metadata. When possible, data centers may try to establish relationships and work procedures with scientists at the beginning of projects. Many researchers are likely to be motivated by short term goals, not long-term goals, which is why, for example, documentation for the long-term is a low priority. Researchers may deposit data and ancillary data (reduced data) into an archive, but have no guarantees that anybody will access these data except the people who deposited them and there may be few rewards even if others do access and use these data.

What might be the "minimum metadata" for a data set? Every domain faces this question. Data citation initiatives also may face an analogous question: What is or ought to be the minimum metadata for data citations? With data sets that are created within large distributed collaborations, identifying the data set "authors" could be difficult. In these cases, a data set may be attributed to a project, or, in analogy to movie credits, individuals may be attributed based on their individual contributions to a project. Establishing minimum metadata standards for data citations, however, could be fairly domain specific, but in most cases they would probably look a lot like the Dublin Core metadata schema: who did it, when did they do it, what is covered, what it is called, how to find it, and the like. This is essentially the Dublin Core "Kernel" metadata. Data citations, however, can only include a small amount of metadata. Extensive descriptions of a data set or the individual contributions to a large project may be best documented in other locations, such as a data set or project's website. The entire data collection level is currently the most common data citation recommendation. Do collection level data citations meet the minimum bar for all disciplines?

Within academic scientific projects, as one participant noted, data work usually defaults to graduate and postdoctoral students. Data loss can thus be a significant issue. When students leave, their data can be lost to the broader project(s) in which they were situated. This can cause trouble in re-creating experiments and is impossible for observations of unique phenomena. This data loss due to losing students is not new, however. This was already the case 30 years ago, and probably even further back. The significance of this issue is that if one cannot reproduce experiments from a lab, one cannot expect anybody else to reproduce them outside of one's own lab. Few papers that are read are actually replicated, however, and then frequently only at substantial cost and effort. Replication also usually only happens if there is a suspicion that something was wrong with the original study.

Another participant said that there may be a need for documented workflows for provenance, but it ought to be easy to generate such provenance documentation or nobody will do it. With regard to "workflow" tools, if they work with a "click" online, then they will probably be used, but otherwise, probably not. If one can take a snapshot of a laboratory via a workflow tool, then there is the problem of distinguishing what is relevant to a particular issue from what is not. Many small steps in a data workflow pipeline may be purely of local interest and not really part of the science. Data reduction—pulling the data relevant to a particular issue out of a larger set of data—is part of the scientific intellectual process.

The question came up as to why it is that most work is not represented in workflows now. Several participants commented that this is probably because most processes do not map to a workflow. Workflows usually work best for repetitive processes. At best, in other scientific work settings, scripts and directories that a student may have left behind become the responsibility of the next person who is hired. Most workflow tools are developed by computer scientists and have not fully penetrated the scientific fields. There are not many examples yet of workflows that have led to scientific breakthroughs. Some scientific projects, within bioinformatics for example, are similar to software projects. In those situations, workflows may be more apt. The most used workflows, such as Taverna, Wings, and Galaxy, are mostly used in computational sciences. Within structural genomics, on the other hand, workflows never really got off the ground, even though some researchers would say that they are using them.

In relation to data citation, workflows are a scientific notebook instantiated via a digital technology. Are publishing and citing these workflows relevant to data citation? If one can make it easy, workflow tools can be a pre-requisite to a data citation, enabling citations to be generated automatically. If a "button" existed to generate data citations out of a workflow plan, scientists might use it, but currently there is no such shortcut. Workflows might also allow some metadata to "fall out" for free. Data centers often find themselves in the situation where the data that are coming in already have lost metadata. If those metadata were captured in workflows, metadata might be maintained more easily. However, workflows might also cause you to lose transparency by "black-boxing" the steps that take place in a data pipeline.

Many participants agreed that scientists need rewards to incentivize data management, and, correspondingly, data citations. If rewards existed, people might do it. For example, writing a research paper is just as time-consuming as documenting data, but scientists write papers all the time because of the rewards given (or at least expected) for publishing them.

Citing data can be difficult, in part, because counting data sets is difficult. For example, one UK research assessment had the option of counting data sets, but not many were available to be counted. This is an issue in the United States as well. The NSF is talking about enabling people to cite data or their contributions to data in their résumés.

Peer review is also an issue here. Are data citable only if they are peer reviewed, or can data be cited as long as they are accessible?

One participant noted that most of our examples of data citation are from *e*-science or big science. A significant exception is the Dryad data repository, which has ecological and evolutionary biology data from small-scale projects. How do scientists in all fields relate to data? Scientists might think of a research idea and then look for data to investigate the idea, or they might see what data are available and develop research ideas that those data can address. "Big sciences" do tend to think about data more, partly because they have to have data management plans ahead of time in order to get funding. Collaborative research versus research that is carried out by individual Principal Investigators shows a dichotomy with regard to data management. There might also be a dichotomy by career stage; for example, university deans might not care about data management because of where they are in their careers. Diane Harley's presentation in an earlier session spoke to this issue more explicitly: Scientists themselves have a responsibility to support data management and citation issues. For example, metrics on data use often are not released, in part because some high profile data sets might not be used much.

Another dichotomy is between disciplinary data repositories and institutional repositories, as one participant observed. Different issues exist for each kind. Institutional repositories tend to be library services. There are few successes with respect to institutional repositories. One reason for this lack of success is that people have to be vested in a repository in order to use it. It can be easier to convince people to submit to a repository if they know that there are people looking in that repository who are relevant to them. Disciplinary repositories may have been more successful for that reason.

"Domain specific repositories," however, can mean "silos of excellence". The question arose as to whether successful multi-disciplinary repositories exist. Even within disciplines, repositories can be problematic. A common issue is that many individual repositories may have been funded within a particular discipline. As services are added on top of those, answers cannot be found within one repository. Instead, the answers are across repositories. Some groups, like the Virtual Observatory in Astronomy are trying to develop services that cut across disciplinary repositories.

As repository connections and consolidation take place, several participants observed that the location of data sets may need to change for the purpose of making scientific research easier. Moving data from repository to repository might make any location binding in a data citation potentially dangerous, although the use of the Digital Object Identifier (DOI) is trying to mitigate this problem. Identification is the first step to linking, but citation is more than linking.

Some of the participants discussed the fact that citations to journal articles perform many functions and that we are trying to shoehorn all of these functions into data citations. Citations create the thread of science. They follow traces of use, both positive and negative uses of a resource. Citations may be in support of or in refutation of a finding. One problem is that researchers rarely publish negative findings. For example, most crystallography research only reports positive results, when, in fact, most research attempts fail. Publishing data, however, might make it easier to cite negative data; that is, data that show negative research results. In structural genomics, you do not get funding if you do not publish negative results in addition to positive ones.

One participant observed that researchers learn over time about how to document research practices and software code. For example, one suggested practice when writing code is to insert tests into code that help to ensure quality. These quality checks, however, typically slow programmers down in the short term. The situation is similar when documenting and working with data. Anecdotally, some scientists report that they spend too much time on working with data, and not enough time doing science. Data citations seem simple, so why is it so hard for people to do them? Bibliographic importing could make this a one-click issue. Creating metadata, however, might be harder. Data centers require a lot of metadata, in some cases perhaps more than the scientific community may be willing to provide.

Another question that was identified in the breakout discussion focused on whether there is any field where data citation may be the norm. Focusing on positive examples might help to illuminate the issue. One example that was raised is in geology. There is a fossil registry that generally everyone uses. They have a specific citation method with hundreds of years of history. This is not *e*-research, however. Fossil resources are not digital. They also have an extensive catalog of single objects, which is not typical in *e*-research. What else is different here? These fossil data come slowly over time as new fossils are discovered. Also, fossils are typically only uncovered via a large time and money investment. Perhaps those resources are seen as having more value because of that investment. As a contrast, in crystallography, it used to take a whole lifetime to develop data sets, but now it has become very easy with digital techniques. Perhaps there is a notion of "canonical" data that applies to fossils.

Other examples were raised by the participants. One concerned sea-surface temperature data held in the data archive of the National Center for Atmospheric Research. Anybody who does research with sea-surface temperature typically uses that data set because it is community developed, comprehensive, and maintained over time. Similarly, the census data are widely used and cited. "Benchmark" data are another example; that is, data that are used to evaluate algorithms in information retrieval or visual image processing research. Behind these canonical data sets are methods, and these citable data sets are seen as "gold standards" for quality data. Some of these canonical data sets are quite old, however, and it sometimes may be useful to update them using new technology. It takes ongoing community development efforts, however, to update such data.

One of the participants noted that in some ways, data citation is a simple problem: provide people with the recommended citation formats and assign data sets DOIs. Why would people still not cite data if these are available? In some cases, it might never occur to people that data have any value, even though they have used them. People may not recognize that they have used somebody else's data, but if you walk them through their data processes, in fact they have used data from other sources. For example, marine biologists forget about tide data, even though tide data are critical to their work. To scientists, asking them to cite data might be like asking them to cite where they got their laboratory chemicals. Data might be seen as a tool, not as an intellectual resource to be cited.

It was noted further that one of the biggest scientific challenges with regard to data citation may be changing the scientific culture so that citing data becomes a regular practice. Scientific practices change gradually, so outreach is useful to implement good data citation practices. The "tipping point" for data citations might not be something obvious. In the United Kingdom, the "freedom of information act" is having an impact, because researchers are more aware that their data may be requested. There is no equivalent requirement that reaches down to National Science Foundation (NSF) grantees. NSF grantees may be awarded exclusive legal protection for their data.

Several participants observed that the funding agencies' data management planning policies might be a lever arm as they evolve. The business models are unclear for data sharing. In some circles (usually people outside of the research process) data are seen as so abundant that they must be easy to share, but this may not be the case. Citation and free access are not directly connected. Access does not imply "no cost." For example, there could be a "fee for service" model in data archives, some cost of a grant would go to cyber-infrastructure that enables data archiving and sharing. When the NSF calls for new big infrastructure proposals, some fields may not respond well because they already have invested in infrastructure independently. There are no single repositories in discipline fields, so cost models might differentiate how users adopt them. The costs for a "fee for service" model of data archiving could be front loaded; the initial users could take the biggest hits because of the small initial user groups. Economies of scale might be slow to grow, as well.

Other questions were raised in this regard. Do fee for service models exist and work? The Inter-university Consortium for Political and Social Research (ICPSR) has a kind of "fee for service" model, specifically, a mixed membership model. Some ICPSR data are only available to members and some data are available to anyone. Can we take lessons from the citation indexing business model? Citation indexing took a few decades to become an accepted tool, and only after Eugene Garfield (and his collaborators) championed these tools in numerous settings and founded a company to enable their work.

One of the participants noted that cloud computing models are gaining traction in scientific fields. Evernote, Basecamp, Dropbox, and others, are widely used cloud-based tools. Anyone can get them and they are very easy to use. Cloud tools are not necessarily interoperable; resources can be siloed in cloud tools just as easily as they can be siloed with conventional technologies. Cloud computing can also introduce a whole new set of confidentiality issues, and is not without some costs as well. With regard to data citation, how do you cite data that are in the cloud? Data may be distributed across computers, and in some cases mirrored or duplicated.

One of the biggest scientific issues related to data citation identified in this breakout session was the culture shift that might be required in many research domains. Currently, citation of data is not widespread in most research communities and is not the accepted thing to do. What is the "tipping point" for data citation? What will push researchers to cite data? One possibility is that a new reward structure for scholarship could be developed. For example, the tipping point for data citations might be when somebody starts counting data set citations. Even if such counting becomes the norm, however, new data citation metrics ought to be developed within institutional structures such as data centers, libraries, universities, and other institutions that provide individual rewards to scientists.

Breakout Session on Institutional, Financial, Legal, and Socio-cultural Issues

Moderator: Vishwas Chavan
Rapporteur: Laura Wynholds

This group faced the challenge of wrapping institutional, financial, legal and socio-cultural issues into a single session. Given that the focus was broad, the conversations branched and circled around the dependencies of data citation. Citation is one aspect of larger systems, such as scholarly communication, academic work, and data archives. Arguably, it lies at the nexus of these established systems, all three of which are in the position of having a considerable installed base as well having practices in flux, so that the outcomes are speculative.

One participant observed that a whole curatorial system is lacking in comprehensively addressing data citation, which some referred to as infrastructure. Others were keen to point out that most people do not include workforce and best practices under the term "infrastructure", both of which are issues here. It was also noted that best practices are a collective responsibility that represent a two-way street between the users and the system.

The following major issues were identified for further discussion:

- resources for infrastructures and human resources for both data and metadata;
- enhancing the recognition for data publication and citation;
- financial sustainability of infrastructure for publishing data and metadata;
- being able to appraise the value of data;
- costs versus benefits of data citation;
- issues of intellectual property (IP), privacy, security, sensitive data, public-private data (confidentiality versus openness); and
- creating a culture of authoring good metadata.

From the outset, it was noted that a single approach for all of these issues is not likely to be effective. However, questions remained, such as what issues would be amenable to a collective approach? In which disciplinary approaches? What are the barriers to uptake? Some of the participants thought that the disincentives to sharing data were paramount. Others felt that culture change around describing and citing data was extremely important. Finally, the issue of appraising the value of data versus the costs of curating it remains.

External Dependencies that Impact Data Citation

In the discussions, there were several major external systems noted that interact with data citation, namely scholarly communication, data sharing, academic work, and data archiving. Moreover, these discussions were so intimately intertwined with data citation, that they were often conflated, such as in the discussion of barriers to data sharing being seen as a barrier to data citation.

1. Scholarly Communication

One participant noted the usefulness of integrating data citation practices with an existing system of scholarly publications, which themselves are used to measure and track scholarly output. There has been increasing awareness of the importance of data publication, and increasing pressure from funders to make research data available. However, while there are a number of models of data publication in existence, the practices are still unstable. Some journals are investing in supporting data in conjunction with the articles, while others are discontinuing the supplemental submissions after a trial period of a few years. Institutions are also acting as publishers via institutional repositories, and have a need to get credit, but they cannot enforce compliance in the same way that journals can. The importance of the disciplinary community defining data citation policies came up again and again. The degree of uptake and implementation varies across disciplines, and cross-disciplinary issues lack attention. It was also noted that getting the buy-in from key editors would be important.

It was posited that currently the transaction costs are too high for data publishing, requiring too much work from too few users. In cases where network effects could be realized from aggregating data, then it could become worthwhile for journals or societies to archive data.

Data citation and publication themselves are metaphors taken from scholarly publication, some participants mentioned. There are tensions around applying print publication models to data, especially since IP rights are different for data and the protections offered vary significantly between countries. Moreover, the law does not match what is being done in practice. In order for the metaphors of data citation and publishing to be useful, it can be useful to understand what it is that we want to count and how it is different from other kinds of publications.

2. Data Sharing

Understanding who shares what data and why is an underlying factor for understanding data citation practices. Christine Borgman's "Conundrum" paper (JASIST, in press?) discusses these incentives and disincentives. It was observed by some participants in the group that there was a fair amount of good will towards sharing across the domains, with comments such as "scholars will share because it is the right thing to do, as long as it is not too much work or too risky" and "every time I share data I learn something". Data sharing is seen as part of moving the field forward, although funding agencies are requiring it as well.

A large part of the discussion on data sharing was airing concerns about disincentives to such sharing. Foremost were concerns expressed about the cost of curating data. A part of this was the observation that not all data are equal, nor should all data be shared. Scientists have a general fear that their data will be misused, misrepresented, misconstrued, or used for purposes that are antithetical to the scientist. One of the discussants noted that there is currently a public relations attack going on about chronic fatigue syndrome that has escalated to threats against personal safety.

Within the issues about data sharing are also concerns about incentivizing data reuse to drive demand. Data intensive fields may have more incentives to reuse data. There are some common issues across many disciplines, but as one approaches the next level of detail, the constraints for data reuse have variation. Libraries do worry about the interdisciplinarity and the cross cutting issues, and then the more discipline-specific concerns.

The first major group of disincentives to data sharing dealt with legal issues and privacy. The legal issues were discussed first. It was asserted that while intellectual property rights have been developed into a maturing system of rights and responsibilities, privacy concerns are still an open problem. With pervasive mobile data collection possible at previously unimaginable scales, privacy has become a significant issue.

It was observed that institutions will have to deal with the privacy concerns posed by data collection or face liability. There are some extant models of privacy around social science survey data, where the publically aggregated data is anonymized, but more detailed data must be accessed via a controlled process. However, it was also noted that many of the privacy issues are separate from data citation.

In addition to privacy issues, there are other access barriers, such as national security, law enforcement, and sensitive data, all of which place limitations on data sharing. Some have seen conflicts arise at the intersections of communities, for example when university faculty collaborated with a certain federal agency, in which the faculty was under a huge pressure to make their data available as soon as they were collected, ignoring the faculty's right to first publication. There was the suggestion that more work needed to be done to set up practices that recognize the rights and responsibilities of individuals and the handling of sensitive data.

Finally, some scientists fear that their data will be misused or used for purposes that are antithetical to their own. Others are concerned that the data can be manipulated to attack the researcher's credibility, as with some of the climate science controversies, or misrepresented to support political agendas. In relation to this need, Creative Commons is working on a standard where any changes to the data are declared within the metadata.

3. Data Archives and Repositories

Data citation is functionally dependent on a storage location for the data. On the surface, data citation is about giving credit for sources used. The persistence of those sources is assumed for purposes of credibility and reproducibility. Ensuring access to a snapshot of the data is expected, both by funders and by publishers, although often not in perpetuity, but rather for a reasonable period of time. A reasonable time frame would present the opportunity for institutions and archives to harvest a copy for safe-keeping.

The question remains as to whether institutions may have a greater role to play in ensuring long-term access to data. In areas where data archives are lacking, some journals have been stepping up to the role of ensuring access and providing storage, such as the Ecological Society of America. Some journals do see that as their role. Some researchers said they have questions as to how long that will last, given the example presented earlier in the day of the journal jettisoning its supplemental materials entirely. Others noted that aiming for "permanently accessible" data was unrealistic, that they would focus the discussion on ensuring access for a reasonable period of time. It was noted by one participant that this sense of a reasonable period of time (rather than in perpetuity) came from the NSF's blue ribbon task force report on sustainability. It was not modeled on the "put it away forever" paradigm, and that material would be moved and reappraised regularly. However, the thing about data is that you will be dealing with more mobile artifacts than the traditionally archival perspective.

Data selection and appraisal were noted as an important feature of data curation with which data citation could assist. It was observed that data curation is in need of better heuristics to inform management decisions. NASA did a study looking for data sets that had never been used, and they discovered that it was about 80 percent of the data they held. These results were somewhat skewed by the fact that NASA keeps multiple versions of some datasets (raw, processed, reprocessed), where the raw data may take up a considerable amount of disk space, but it is the processed versions that receive the majority of use. Libraries largely operate on a model of collective action that is based on redundancy, whereas data archives tend to hold data that has no redundancy, and thus the archival paradigm may be more appropriate for modeling selection and appraisal decisions.

4. *Academic Work and Workforce*

In some ways, the larger question remains as to who is going to do this work of ensuring access and availability to research data. Creating a data curation workforce is an open discussion in the information science world. Education and challenges remain, but another aspect is funding the work that employs the practitioners. Some of the session participants raised questions such as, Should the work be done within the library? Or, should the work be done using embedded digital curation team members?

Such workforce issues will be important to consider as we move into a data intensive paradigm of science. A professional class may need to be supported to make the data accessible, citable, and persistent.

5. *Challenges of Establishing the Value of Academic Data*

One of the participants noted that there are difficulties with establishing the value of data citation is that it is also related to valuation of the science possible with the data. It is considered more valuable if the data supports new science as opposed to incremental science. There can be a prejudice against reuse because it is not considered as captivating as doing new science. It was explained in terms of being worthy of a Nobel Prize: if the data reuse is not Nobel worthy, then it is really hard to attract good scientists to it. For scientists who work with reusing data in this paradigm, they are considered to be giving up their careers.

The NSF is starting to try to incentivize data reuse, as seen with the new funding opportunity from NSF for reusing certain types of data. The United Kingdom may also see some movement on this front with some of the legislation pending in Parliament.

Much of the discussion about building credit for data producers was driven from the perspective of the tenure and promotion process in academia. The role of providing credit and the system of rewards are both different for those outside the "publish or perish" system of academia.

It was pointed out that another major stakeholder in data citation is the data center. As an organization whose mission is to produce data, as opposed to a professor in an academic perspective, the situation is quite different. The data itself are the end product. The NASA Earth Observing System Data and Information System (EOSDIS) is an example of this. All of this is important for the advancement of science. It would be interesting to consider what fraction of the data we are discussing comes from different types of sources.

There is a lot of interest in the ways that a given set of data is cited in the peer reviewed literature. For example, there is a study to look at the scholarly impact of one of the instruments on the NASA EOS satellites. A challenge with that is that there are important uses of data outside of the scholarly world. One of the presentations earlier in this workshop demonstrated an example of a citation of a data set in a study by a non-governmental organization of a proposed reforestation project. What is the value of that particular citation (which is not in the peer-reviewed literature)?

As the presentation by Bruce Wilson indicated, the ORNL DAAC cares about citations—partly because it ties back to giving credit to the people who provide us the data. It also reflects back on the value of the ORNL DAAC as a data center, providing the role of that cadre of people who are doing the curation, discussed in Diane Harley's talk. ORNL needs those citations and use metrics as a means to (a) understand what data are being used and why; that is, how to lower the barriers to the use of the data; and (b) justify their budget.

Models of Data Citation

Some of the participants suggested that the current practices of citing data have yet to coalesce around best practices and standards, so there are outstanding questions about how data citation fits in with data sharing and data publishing. Citation as a scholarly practice and the citation of data within it present a variety of models for best practices. As discussed above, data publication itself presents challenges, but it was seen by many participants as central to getting data citation off the ground. It was also noted that there was likely no single solution to these challenges. Of central importance to data citation was the intention to build credibility for creators throughout the lifecycle, but there were also technological dependencies around cost and ease of use. The discussion was two-fold. On the one hand were concerns about what information was necessary for citation purposes, on the other was the question of how to leverage data citation.

Within the discussion of data citation models and standards, much of the concern was about fulfilling the functions performed by data citation. The functions of citation were not explicitly enumerated, but among those discussed were tracking usage via citation metrics, transaction costs and overhead for tracking usage, and whether citation standards impact the cost of implementation. It was noted that there was some tension between repositories that leaned toward including more elements within the citation, enumerating responsible parties and agencies, and the publishers who preferred a shorter, simpler template with as few elements as possible. There is also the tension of academic institutions as employers, but also as providers of services to other institutions and persons.

There was also a fair amount of discussion around compliance, norms, and how to impose a mechanism such as data citation. One way to do that would be to have an open standard that has agreed upon elements, but the journals want short citations and the repositories want longer ones to help attract funders. There was some question about what the minimum number of elements would be to satisfy the stakeholders: publishers, institutions, CrossRef, DataCite, data authors, and so on. There was also some discussion about whether a human readable name was necessary.

There was some discussion about whether it would be feasible to embed metadata in the resolved page. HTML was the example cited, using a simple, weak, extensible protocol, such as a landing page with all the necessary credits. There was the concern that having the DOI resolve the full metadata. However, there also was concern that this type of approach would be too brittle for the long term.

1. Current Approaches

Current approaches to data citation largely follow disciplinary practices. It was suggested that one approach would be to agree on the purpose of citation, track the mechanisms, and see how they work for different disciplines. There was some concern over the splintering of standards across disciplines with that kind of approach. It was also pointed out that some disciplines have functioning practices already in place and whatever is implemented should not force changes on that which already works.

However, as we have seen in this workshop, there is DataCite, which is interdisciplinary and largely a library organization. The question was raised of how is DataCite going to expand and do what it wants to do? It was noted that they are in collaboration with CrossRef and reaching out to the publishing community. CrossRef is largely focusing on more traditional document type of publications and DataCite is focusing on data. Thomson Reuters would like to start indexing datasets and including them in their Web of Science. Much of this activity is focusing on the sciences (rather than the humanities). It was also noted that many of the data centers achieved buy-in from publishers by using DOIs, as it leverages the reputations and workflows of these identifiers. It was observed that there is some tension in aligning needs within DataCite as the UC3 and Purdue partners are the only academic institutions, with the rest being national libraries.

Dryad, for example, makes its data available under a Creative Commons Zero (CC0) license, which receives a fair amount of resistance from depositors. CC0, much like traditional citation practices, relies on norms of scholarship, rather than on legally binding contractual language. It was observed that CCO does not naturally port very well to scholarship and data and presents the potential to yield unintended consequences.

The American Geophysical Union (AGU) requires its journal authors to cite data and open their data by placing them in a data center. They also limit the citing of datasets, stipulating that one cannot cite datasets that are not permanently archived, but rather such data must be acknowledged like a personal communication. In this case, the term "cite" is a term of art. The AGU is not unique in this regard. There are a number of journals that follow this model because of the discoverability and access issues that non-archived materials present. In these cases, citations are used for formal audited sources; acknowledgements are for less formal sources. It was observed that this kind of stratification of sources could serve as a selection process for institutional repositories ingesting materials.

2. Identity, Data Structure, and Provenance

Identity and provenance are known challenges to both data citation and data sharing. These issues were brought up by the participants under concerns about taking subsets of data and ensuring reproducibility. How does the user know if they are accessing the same data? These issues of reproducibility assume that the data are static, but they also assume a stable repository. There are many examples of researchers taking raw data and manipulating them to such an extent that it is questionable whether they should even be considered the same data.

Identifiers were seen as a central feature of discoverability and access. The costs of registering DOIs with CrossRef and DataCite were discussed, as well as some of the indexing services that make use of the metadata.

Some participants brought up the model of mandatory copyright deposit for national libraries in Europe, as at the U.S. Library of Congress as a possible model for data curation, because it allows users and institutions to request copies. However, the Library of Congress has already decided that data are, generally speaking, outside of their scope. Within that model is the assumption that what is taken is kept in perpetuity, which is a huge economic issue.

3. Costs

The question of how to determine the value of data citation was pondered, getting into the incentives and benefits of having the data cited. The cost of data citation is complex. On the one side is the cost of labor of creating the citation and the cost of minting the identifier. On the other side is the cost saving in labor in discoverability. There is also the cost of doing nothing and having the data be very difficult to find and access. Thomson Reuters, Ebsco, and others are watching databases with an eye towards being able to start indexing them in their services. The cost of the identifiers specifically and data curation more generally was difficult to assess, given the potential cost savings in creating an economy of scale around data curation and discovery for scientists. It was advocated that the cost of the infrastructure and human resources was relatively small compared to the benefits. However, that assertion would need to be quantified, with the question of how data are being used is still outstanding. Some studies suggest that the investments in data curation pay for themselves.

Other costs discussed by the group included the wasted opportunity costs of reinventing the wheel and redoing the same research because researchers were unaware that the research had already been done, the cost of redundant studies that place people and animals at risk, the cost of toxic experiments such as nuclear experiments, many of which have been accomplished by resifting old data.

Leveraging Adoption

A number of participants observed that given the complexity of the situation, leverage may be needed to encourage adoption, to change mindsets, and to change what is valued. Examples of this have included setting examples of good practices for the younger individuals, collective value and emphasis - "it starts in your lab" (we need posters). Data citation has a relationship to the role of credit for different stakeholders. In creating a culture of making data available and citing them, one also has to create a culture of valuing the data such that they can be considered relevant for tenure and promotion decisions. However, it is also about valuing and rewarding reuse, as well as enabling reproducibility.

Several participants remarked that behavior change was part of what was needed. Diane Harley's presentation noted that this happens through changes in expectations. There is a substantial literature on social changes, including the literature of technology adoption. There were questions about what can be learned from the literature that is relevant to this particular discussion. Does it tie back to the earlier comments that the science work itself is potentially the subject of future work on the history and development of science?

Later discussions also pondered the importance of the policy environment, policy authorship, and policy compliance in data citation. There was also some discussion of who was responsible for setting best practices, and domain specific versus institution specific policies and practices. It was noted that since publishing practices center around disciplinary communities, data citation policies will also need to define data citation practices in different disciplines.

Many venues mandate that researchers must cite the data that they use as the result of what researchers are obligated to do when they receive data. In some cases, usage licenses are being written. (See also Sarah Callaghan's example mandating data citation.) The question remains as to whether institutions should try to control citation via usage licenses which can demonstrate impact for the data's expense. On the one hand, institutions need to provide clear best practices for their researchers, but on the other hand, they also need to provide compliance with using third-party data. Some questioned whether any of these approaches were realistic for faculty to adopt.

Finally, others asserted that it really depends on the datasets and the use. If the dataset is used, one should be able to cite it. If someone spends two years collecting data, however, it is generally accepted that they can use them exclusively and not share them for a period of time.

1. *The Importance of Disciplinary Norms*

Mirroring other discussions, the importance of the disciplinary community was key to this discussion as well. Some in the breakout group noted that the disciplinary community has the power to instruct their constituents to cite data. One observation indicated that the pattern of data sharing was of small communities coming together to share, then getting approval from the journal editors. Consensus for how data are to be cited has to be built at the disciplinary level.

It was postulated that the publishers need to abide by the culture of describing their datasets well with voluntary compliance to citation. The notion of credit is also important for data citation. It was observed that many researchers seem to have stories about data citation problems and receiving credit, but that these problems do not seem to get addressed. In this area, the normative aspects do not seem to be as normative.

2. *Funding Mandates*

Funding mandates for citation and access was another major discussion point, especially with the recent discourse concerning data management plans.

3. Tracking Use

Citation is important as a scholarly activity because it provides a way to follow usage for people who contribute data. Citation is an incentive in that way. It was asserted that the need for a system goes far beyond a citation, however. There is currently little incentive to cite data because netting citations to data is not considered for most academic tenure and promotion decisions. Usage data has become quite important in other areas of scholarship and leads to impacts beyond the initial usage data. It raises the question as to whether when you cite a dataset you become complicit in the future funding. There were examples from business schools, where they were charging for use and access to their datasets.

One question, partially discussed above, is whether the data work and resulting citation will rely on goodwill and norms for compliance or licenses that carry specific obligations. It was asserted that citation should have opt-out mechanisms that are trackable. That way you can discipline non-compliance. It was also observed that there is some tension between licenses and norms, with licenses having the potential to yield significant unintended consequences. If we could say that norms would be dominant, then we could talk scientists out of licenses, but we are seeing more licenses rather than fewer. It was also observed that if data are in a standardized structure, then it becomes advantageous to archive, as we see with the American Chemical Society.

Unfortunately, citation is a very lagging indicator of use. ORNL sees an average of 18-24 months from when the data are downloaded until they see a citation in the literature. A related example of barriers includes work that the ORNL DAAC did to make some data available via OGC web services. Data download rates increased on the order of 100 times for some of those data sets by making them available by OGC web services. Part of this effect was caused by advertising. They are now starting to see some of that increase in downloads show up in citation rates. It is not 100 times, but the increase may well be significant.

The main concern for data centers is demonstrating use of their data; NASA archives have a senior review every 2-3 years, and if they cannot show that their data are being used for peer-reviewed publications, their funding gets cut, and the data might go offline or to a less costly storage and management system. Missions have similar reviews once past their primary mission schedule, and if people are not using the data, the mission is terminated. Although they also track download volume, this can be a bad metric, as the network bandwidth rates are not keeping up. The data centers are making a concerted effort to save people's time by facilitating more targeted downloads of data (e.g., reduced or lower cadence data to identify the periods and locations of interest, then serve subsets of the data rather than the full dataset).

4. Accountability and Transparency

As mentioned above, citation provides a mechanism for tracking use. Conversely, it also is a way to establish that you have shared your data. It provides a mechanism for accountability and transparency. At the IPCC, the notion of accountability has come to the fore, derived from false accusations of impropriety, but is being used to develop better transparency.

However, this sense of accountability and transparency is not entirely an incentive. There are also concerns about data being scooped or stolen, with junior faculty and postdoctoral researchers being particularly vulnerable. One participant cited an example of a colleague in Los Angeles who was interviewed by the press, but was not able to open his data as of yet due to a pending publication in *Nature*. He received a nasty editorial in the press for being a public employee and not sharing his data.

5. *Embargos and Proprietary Periods for Data*

Embargos, which are also referred to as proprietary periods of exclusive use in some fields, are generally seen as an important tool for protecting data, for protecting postdoctoral researchers and junior faculty, for protecting dissertation work until the derivative publications are finished, and generally for maintaining the primacy of researchers. Conversely, data registries are generally not on the researchers' radars.

The Long Term Ecological Research Network has done some work on this, for example. There is some view that having an embargo set up at the time of deposit, with a particular sunset date, is best. The idea is that the embargo should be a standard length of time (e.g., 2 years). The researcher can extend the embargo, but the embargo will automatically end without an express action.

Breakout Session on Institutional Roles and Perspectives

Moderator: Bonnie Carroll
Rapporteur: Jillian Wallis

Several participants began by focusing on the stakeholders and low-level details about the interaction between the stakeholders and the data citations. Others then raised several questions: Who is cited: the data center hosting the data, the data producer, or anyone who has added value to the data? This is really a question of whether the citation is for assigning credit or finding data. It should be noted that there are many stakeholders who add value to the data and it may not be feasible to acknowledge everyone. Who is responsible for generating a citation: the data center hosting the data, some collaboration between the producer and archivist, or the data user consulting with the data producer to create a citation? The credit aspects of citation thus may conflict with the location and discoverability aspects, which have very different sets of requirements.

A number of the participants identified issues that pulled apart the roles of data citation stakeholders. Who should be the citation creator: the data creator responsible for providing a citable thing, or the data user responsible for citing that thing? Who is responsible for collecting metrics? This led to plotting out the events that happen during the life of a data citation and assigning responsible parties. Figure S-1 presents one understanding of how data citations will come to be. Rather than being a representation of the life-cycle of an individual data citation, it instead depicts the life cycle of how data citation practices in general will be created. In this case, life-cycle is perhaps a misnomer, and instead what is captured in the figure is a timeline for organizing all of the interested parties.

It is important to further define the data citation lifecycle and the roles and responsibilities of institutions and people who act at each stage, in order to determine who is missing from this discussion and how we can get them involved.

Event	Responsible Party
Understanding/frameworks	Universities (Academia)
Policy	Funders Universities Journals
Define Elements	Community
Layout	Standards Bodies Publishers
Define Content	Data Set Creators Data Centers
Create Citation	Data User
Train/Educate	Libraries
*Aggregation	Commercial Citation User

Birth of a citation brackets: Define Elements, Layout, Define Content, Create Citation. Derivative Datacycle loops from Define Content.

FIGURE S-1 Data citation lifecycle.

Prior to the actual creation and adoption of data citations, several participants suggested, one option is to develop an understanding of the social ramifications of the data citation and the frameworks with which data citations would need to interact. This understanding could come from academic research on data practices. At the top level, research funders, universities, and journal publishers could think about developing a data citation policy that supports their respective needs and creates incentives to encourage data citation.

Using such a base of understanding and policy, many parties may wish to work in parallel to make data citation a reality. Research communities can define the data citation elements that are meaningful to them. Journal publishers and standards bodies can define general data citation layouts that are both machine and human-readable. In order for a data citation to be created: (i) the data need to have been generated by someone, and (ii) the data need to be available with enough information attached in order to create the data citation. The data generator or the data center hosting the data will then make the actual citation content available. The data users are responsible for actually using the data citation in their publications. The derivative data cycle here refers to the practice of creating derivative datasets from other datasets. A new form of data citation could be developed in order to take this practice into account, and can involve some combination of the original data generators or hosts and the data users in a new data citation or a data citation that expands into multiple data citations.

Once the various standards are in play, several participants remarked that training and education would be useful about how and when data citations can be used. The university libraries are perhaps well positioned to reach out to the academic communities they support. Finally, commercial parties can aggregate data citations, much like citations are aggregated to characterize scholarly communication in the literature.

Appendix A: Agenda

Developing Data Attribution and Citation Practices and Standards
An International Symposium and Workshop
August 22-23, 2011

US CODATA and the Board on Research Data and Information
in collaboration with
CODATA-ICSTI Task Group on Data Citation Standards and Practices

AGENDA

Day One – Monday, August 22

9:00 am — I. Chair's Welcoming Remarks and Keynote: Why are the attribution and citation of scientific data important?

    Christine Borgman, University of California at Los Angeles

9:20 — II.a. What are the major technical issues that need to be considered in developing and implementing scientific data citation standards and practices?

    Moderator: John Wilbanks, Creative Commons

1. How attribution and citation relate or differ: Jean-Bernard Minster, University of California at San Diego, Scripps Institution of Oceanography

2. Attribution and Credit: Johan Bollen, Indiana University

3. Persistence, identification, and the actionability of data citations:

Herbert van de Sompel, Los Alamos National Laboratory

4. Authenticity, provenance, and trust - maintaining the scholarly value chain: Paul Groth, VU University Amsterdam, Netherlands

- Discussion

10:50 Break – 30 min

11:20 II.b. What are the major scientific issues that need to be considered in developing and implementing scientific data citation standards and practices? Which ones are universal for all types of research and which ones are field- or context- specific?

    Moderator: Herbert van de Sompel, LANL

    1. Life Sciences: Philip Bourne, University of California at San Diego

    2. Physical and earth sciences: Sarah Callaghan, Rutherford Appleton Laboratory, UK

    3. Social Sciences: Mary Vardigan, University of Michigan, Inter-university Consortium for Political and Social Research

    4. Humanities: Michael Sperberg-McQueen, Black Mesa Technologies

    - Discussion

12:50 Lunch (70 min, on site)

2:00 III. What are the major institutional, financial, legal, and socio-cultural issues that need to be considered in developing and implementing scientific data citation standards and practices? Which ones are universal for all types of research and which ones are field- or context-specific?

    Moderator: Paul Uhlir, National Research Council

    1. Legal issues: Sarah Hinchliff Pearson, Creative Commons

    2. Institutional/financial: MacKenzie Smith, MIT

    3. Socio-cultural: Diane Harley, University of California at Berkeley

    - Discussion

3:15 Coffee break – 30 min

3:45 IV. What is the status of data attribution and citation practices in individual fields in the natural and social (economic and political) sciences in United States and internationally? Case Studies.

Moderator: David Kochalko, ThomsonReuters

1. DataCite: Jan Brase, National Library of Science and Technology, Germany

2. Dataverse: Micah Altman, Harvard University

3. Microsoft Academic Search: Lee Dirks, Microsoft Research

4. International Oceanographic Data Exchange and the Scientific Committee for Oceanographic Research: Roy Lowry et al. (presentation given by Sarah Callaghan)

5. Global Biodiversity Information Facility: Vishwas Chavan, GBIF

6. Federation of Earth Science Information Partners: Mark Parsons, National Snow and Ice Data Center

7. Scripps Institution of Oceanography: John Helly, Scripps

8. SageCite: Monica Duke, University of Bath, UKOLN

- Discussion

5:30   Adjourn -- reception

Day Two – Tuesday, August 23

Hotel Shattuck Plaza
Whitecotton Room, Sixth Floor
2086 Allston Way
Berkeley, CA

8:45   V. Institutional Roles and Perspectives:

What are the respective roles and approaches of the main actors in the research enterprise and what are the similarities and differences in disciplines and countries? The roles of research funders, universities, data centers, libraries, scientific societies, and publishers will be explored.

   Moderator: Bonnie Carroll, Information International Associates

   1. Universities: Deborah Crawford, Drexel University

   2. Data centers – Bruce Wilson, Oak Ridge National Laboratory

   3. Libraries: Michael Witt, Purdue/IASSIST

   4. Commercial scientific publisher: Anita de Waard, Elsevier Labs

   5. Scientific society publisher: Michael Kurtz, Harvard-Smithsonian Center for Astrophysics, Astrophysics Data System

   - Discussion

10:30   Break (30 minutes)

11:00   Session V. (continued)

   Moderator: Christine Borgman, UCLA

   6. Standards: Todd Carpenter, National Information Standards Organization

   7. Public research funder: Sylvia Spengler, National Science Foundation

   - Discussion and wrap up

12:15	Lunch (1 hour)

Workshop – Options on where do we go from here?
Whitecotton Room, Sixth Floor

Moderator: Allen Renear, University of Illinois at Urbana-Champaign

1:15-1:25	Introduction and charge to breakout groups, Allen Renear, University of Illinois at Urbana-Champaign
1:30 -3:30	**Breakout Groups -** 6 groups @ 7-9 persons each, with moderator and rapporteur (meeting rooms to be assigned)

Breakout 1: Why is the attribution and citation of scientific data important and for what types of data? Is there substantial variation among disciplines?

Chair: Jan Brase, TBI and DataCite, Germany

Rapporteur: Cheryl Levey, NRC Board on Research Data and Information

Room: Boiler Room, Section A

Breakout 2: What are the major technical issues that need to be considered in developing and implementing scientific data citation standards and practices?

Chair: Martie van Deventer, Council for Scientific and Industrial Research, South Africa

Rapporteur: Franciel Linares, Information International Associates

Room: Boiler Room, Section B

Breakout 3: What are the major scientific issues that need to be considered in developing and implementing scientific data citation standards and practices? Which ones are universal for all types of research and which ones are field- or context-specific?

Chair: Sarah Callaghan, Rutherford Appleton Laboratory, UK

Rapporteur: Matthew Mayernik, National Center for Atmospheric Research

Room: Boiler Room, Section C

Breakout 4: What are the major institutional, financial, legal, and socio-cultural issues that need to be considered in developing and implementing scientific data citation standards and practices? Which ones are universal for all types of research and

which ones are field- or context-specific?

Chair: Vishwas Chavan, Global Biodiversity Information Facility, Denmark

Rapporteur: Laura Wynholds, UCLA

Room: Crystal Ballroom, Section 1

Breakout 5: What are some of the options for the successful development and implementation of scientific data citation practices and standards, both across the natural and social sciences and in major contexts of research? How can the different stakeholder groups be engaged in such a process?

Chair: Bonnie Carroll, Information International Associates, US

Rapporteur: Jillian Wallis, UCLA

Main Room, Side 1

Breakout 6: What issues would be useful to get additional feedback on from the scientific community in order to identify best practices for data citation practices and standards? Who should be asked? What is the best way to get this information?

Chair: Todd Carpenter, National Information Standards Organization, USRapporteur: Daniel Cohen, Library of Congress/NRC Board on Research Data and Information

Main Room, Side 2

3:30	Break
4:00	Plenary discussion of best practices and options, and wrap-up
	Chair: Allen Renear, UIUC
5:00	End of meeting

Appendix B: Speaker and Moderator Biographical Information

Speaker	Affiliation	URL for Biographical Information
Micah Altman	Harvard University	http://mit.academia.edu/MicahAltman
Johan Bollen	Indiana University	http://informatics.indiana.edu/jbollen/Home.html
Christine Borgman	University of California at Los Angeles	http://polaris.gseis.ucla.edu/cborgman/Chriss_Site/Welcome.html
Philip Bourne	University of California at San Diego	http://www.sdsc.edu/~bourne/
Jan Brase	National Library of Science and Technology, Germany	http://sites.nationalacademies.org/PGA/brdi/PGA_064146
Sarah Callaghan	Rutherford Appleton Laboratory, UK	http://sites.nationalacademies.org/PGA/brdi/PGA_064138
Todd Carpenter	National Information Standards Organization	http://www.niso.org/about/directory/staff
Bonnie Carroll	Information International Associates	http://www.codata.org/codata02/bios/bio-carroll.htm
Vishwas Chavan	GBIF	http://vishwaschavan.in/
Deborah Crawford	Drexel University	http://sites.nationalacademies.org/PGA/brdi/PGA_064137
Anita de Waard	Elsevier Labs	http://sites.nationalacademies.org/PGA/brdi/PGA_064139
Lee Dirks	Microsoft Research	http://sites.nationalacademies.org/PGA/brdi/PGA_064136
Monica Duke	University of Bath, UKOLN	http://sites.nationalacademies.org/PGA/brdi/PGA_064142
Paul Groth	VU University Amsterdam, Netherlands	http://www.few.vu.nl/~pgroth/Site/Welcome.html

Diane Harley	University of California at Berkeley	http://sites.nationalacademies.org/PGA/brdi/PGA_064151
John Helly	Scripps Institution of Oceanographic Research	http://www.sdsc.edu/profile/jhelly.html
David Kochalko	Thomson Reuters	http://www.stm-assoc.org/people/dave-kochalko/
Michael Kurtz	Harvard-Smithsonian Center for Astrophysics	https://www.cfa.harvard.edu/~kurtz/
Roy Lowry	Plymouth University, UK	http://www.plymouth.ac.uk/staff/rlowry
Jean-Bernard Minster	University of California at San Diego, Scripps Institution of Oceanography	http://www.sio.ucsd.edu/Profile/jbminster
Mark Parsons	National Snow and Ice Data Center	http://sites.nationalacademies.org/PGA/brdi/PGA_064147
Sarah Hinchliff Pearson	Creative Commons	http://sites.nationalacademies.org/PGA/brdi/PGA_064152
Allen Renear	University of Illinois at Urbana-Champaign	http://people.lis.illinois.edu/~renear/renearcv.html
MacKenzie Smith	MIT	http://sites.nationalacademies.org/PGA/brdi/PGA_064149
Sylvia Spengler	National Science Foundation	http://www.nsf.gov/staff/staff_bio.jsp?lan=sspengle&org=CISE
Michael Sperberg-McQueen	Black Mesa Technologies	http://sites.nationalacademies.org/PGA/brdi/PGA_064153
Paul Uhlir	National Research Council	http://sites.nationalacademies.org/PGA/brdi/PGA_059692

Herbert van de Sompel	Los Alamos National Laboratory	http://public.lanl.gov/herbertv/home/
Mary Vardigan	University of Michigan	http://www.icpsr.umich.edu/icpsrweb/shared/ICPSR/staff/vardigan
John Wilbanks	Creative Commons	http://sciencecommons.org/about/whoweare/wilbanks/
Bruce Wilson	Oak Ridge National Laboratory	http://sites.nationalacademies.org/PGA/brdi/PGA_064145
Michael Witt	Purdue/IASSIST	http://sites.nationalacademies.org/PGA/brdi/PGA_064140